Reliability in Biomechanics

Reliability of Multiphysical Systems Set

coordinated by
Abdelkhalak El Hami

Volume 3

Reliability in Biomechanics

Ghias Kharmanda
Abdelkhalak El Hami

iSTE

WILEY

First published 2016 in Great Britain and the United States by ISTE Ltd and John Wiley & Sons, Inc.

ISTE Ltd
27-37 St George's Road
London SW19 4EU
UK

www.iste.co.uk

John Wiley & Sons, Inc.
111 River Street
Hoboken, NJ 07030
USA

www.wiley.com

Library of Congress Control Number: 2016952173

British Library Cataloguing-in-Publication Data
A CIP record for this book is available from the British Library
ISBN 978-1-78630-024-9

Contents

Preface

In the deterministic method, all parameters that have an uncertain nature are represented by unfavorable characteristic values, associated with the safety factor. The deterministic method uses a pessimistic margin determined by the consequences of a probable failure. Most of the time, this method leads to unjustified specifications, in particular for sensitive structures. Reliability analysis has become an invaluable tool for certain sectors such as the nuclear, space, aeronautical and automobile sectors. Failure phenomena must be addressed in detail in these sectors. We can distinguish between three types of studies:

– the reliability of structures;

– the reliability of systems;

– the reliability of results.

Faced with the deterministic method's inability to take into account the diversity of physical phenomena that apply to structures, engineers have developed another method, better-suited to the uncertain nature of physical phenomena. In this method, the failure of a structure occurs if the probability of failure is greater than a preset threshold. This method is called the "probabilistic method".

The probabilistic method is increasingly used in engineering as demonstrated by its different applications in industry. It is used to check that structures of known geometry have an acceptable level of probability, or to optimize the dimensions of a structure in order to respect certain fixed objectives, such as desired cost or the desired level of probability.

Beyond this, reliability analysis is an important tool for making decisions to establish a maintenance and inspection plan. It can also be used for the validation of standards and regulations. To perform reliability analysis, several methods can be used to give the probability of failure in an efficient and simple manner.

This work focuses on the tools necessary for integrating the concept of reliability into biomechanical applications, in particular the design of orthopedic and orthodontic prostheses. We are interested in the reliability of structures to integrate this into biomechanical applications, considering the uncertainty in loading, geometry and materials. Next, the reliability of systems is addressed considering several failure scenarios. The materials of living tissue are very difficult to deal with. For this, formulations with a high level of precision are presented. Finally, several recent methods are addressed to perform reliability analysis for biological applications, in particular orthopedic and orthodontic prostheses.

Acknowledgments

We would like to thank all of those people who have, in some way, great or small, contributed to the writing of this book – in particular, Sophie Le Cann, a researcher at the Biomedical Centre (BMC) at Lund University, for her contribution in terms of biological language. Heartfelt thanks go to our families, to our students, and to our colleagues for the massive moral support whilst this book was being written.

Ghias KHARMANDA
Abdelkhalak EL HAMI
August 2016

Introduction

Reliability analysis is a vast domain that contributes to understanding, modeling and forecasting the degradation and aging mechanisms liable to lead a component to failure and a system to breakdown. Understanding relationships between physical limitations, intrinsic faults, technological imperfections and environmental and internal constraint is the subject of this vast and complex activity. While it is not yet a subject we fully understand, this is something we must work toward. Reliability analysis can still lead to efficient solutions: adapting constraints to physical limitations, protecting the component from internal and external damage or conversely prompting the development of components to make them more robust regarding real constraints.

Research in biomechanics uses approaches in geometric and mechanical modeling, and experimental analysis similar to structural and material mechanics. However, there are numerous obstacles at all stages of characterization and personalization of the geometry, the materials whose behavior varies according to still poorly understood remodeling laws (such as growth, aging, etc.) as well as “in-service” mechanical loading. These key points are also relevant for the validation of models that are as mechanical as they are physiological. Such models require the development of quantitative methods of investigation of living things beyond the traditional domain of mechanics and anatomy, taking into account very strong clinical and ethical constraints. In this book, integrating the concepts of reliability is performed with the aim of dealing with the uncertainty of several aspects, including the loading and the properties of bone materials. The description of the behavior of biological tissue, and bone in particular, is very complex and remains controversial; even more so accurately, changing geometrical parameters during the design of prostheses. In the context of loading, understanding damage mechanisms can contribute to the design of prostheses, as well as improving them. Additionally, the mechanical properties of bone can change after surgery. First, bone is a living "material" whose characteristics vary over time, the type of bone considered, the sex of the individual, their age and health, etc. Furthermore, the modification of mechanical actions, induced, for example, by the installation of an implant, can provoke a radical change in its quality. Theories of bone remodeling and adaptation try to describe and predict its variation over time. Reliability analysis is applied to orthopedic prostheses (hip, knee, etc.) in order to evaluate their level of reliability and stabilization.

The effect of modeling choices on subsequent results has considerable consequences. Taking into account the anisotropy of the bone seems to be crucial in maxillary studies. The bone can be locally considered as an orthotropic material with two orthogonal planes of material symmetry. In a coordinate system linked to these two planes, its elastic behavior is thus defined with the help of nine constants or elastic moduli. However, this coordinate system can change from one point to another in an anatomical part such as the mandible, by orienting itself in a manner to guarantee itself maximum mechanical performance with respect to the usual loading to which it is subjected. A change in this loading due to, for example, the implantation of a prosthesis can lead to a reorientation of this orthotropic coordinate system. Reliability analysis is next applied to evaluate the stability of fixation systems used for mandible fractures.

This work is made up of five chapters:

The *first chapter* presents several basic tools for reliability analysis. Reliability is calculated by a specific optimization procedure. The optimization depends on the sensitivity analysis that determines the influence of all the input parameters on the output parameters. Furthermore, nothing in biomechanics can be described by analytic geometry. For this, the intervention of numerical simulation is an essential stage of solving complex problems. In order to estimate, with high precision, all these mechanical values and determine the reliability of a given solution, the finite element method appears to be a preferred numerical simulation tool. Thus, in this chapter, numerical simulation by finite elements is first presented. Then, a detailed presentation is given for optimization technology (classification and methods). Finally, the sensitivity analysis is presented as an indispensable tool for optimization and can also be used for the reliability in the following chapters.

The *second chapter* is dedicated to basic concepts of reliability. First, the history of reliability concept is presented to include the development of this concept over time. Basic statistical functions such as the probability density function and the cumulative distribution function are then examined along with the necessary statistical measures like the mean and standard deviation. Then, the reliability of systems is presented to include the case of multiple failure scenarios. Several applications of distribution laws are examined, in particular the uniform, normal and log-normal distributions. The reliability analysis approach is finally presented including optimization technology.

The *third chapter* is dedicated to the integration of reliability analysis in biomechanics at several levels: loading, geometry and materials. Then, the uncertainty of bones is examined along with the development of formulations relating the mechanical properties of bone for isotropic and orthotropic behavior. Different works are performed at two levels: macrostructure and microstructure. Experimental works were performed at the solids mechanics laboratory at the University of Lund to validate the developed formulation and extend it to dynamic cases.

The *fourth chapter* is based upon integrating reliability analysis into orthopedic prostheses to evaluate stabilization after surgical operation. An introduction to orthopedic prostheses is given. Then, two types of prostheses are examined to perform the reliability analysis. The first prosthesis studied is the intervertebral disk. An anatomical presentation of the lumbar region of the spine is given followed by a simple numerical treatment of the disk to show the application of reliability in a very simple manner by considering a problem of two geometrical variables. Next, a detailed study of the hip prosthesis is performed starting with an anatomical presentation of the hip. Finally, a strategy is presented for integrating reliability analysis into the hip prosthesis considering the uncertainty of the mechanical properties of the materials used. In this chapter, the numerical treatment is performed in two-dimensional in order to complete the reliability analysis in a simple and pedagogical manner.

The *fifth chapter* demonstrates the integration of reliability analysis into orthodontic prostheses to evaluate stabilization after surgical operation. The strategy of fixation with mini-plates is used for the treatment of fractured mandibles. An anatomical presentation of the mandible is given to understand how the forces of the muscles function. During a surgical operation, the muscles can be damaged or cut. For this, the uncertainty is manifested over the loading by considering the biting force and the muscular forces as random variables in this chapter. An application of numerical simulation over an unfractured mandible is presented to demonstrate the positive role of muscles that is considered during the design. Additionally, reliability analysis is applied to the fractured mandible considering a structure of an isotropic, homogeneous material. Finally, reliability analysis is applied to the fractured mandible considering a structure composed of two orthotropic tissues. In this chapter, the numerical applications are performed in three-dimensions over a clinical case to carry out a more realistic reliability analysis and compare with the surgeon's observations.

1

Basic Tools for Reliability Analysis

1.1. Introduction

The modeling of a mechanical system can be defined as the mathematical idealization of physical phenomena that control it. This obviously requires us to define input variables (geometric parameters of the system, loading conditions, etc.) and output variables (displacements, stresses, etc.) that will help to understand the evolution of the mechanical system. The models used are more complex and accurate and the current issue is the identification of the parameters constituting them. In fact, we can no longer afford, in dealing with certain types of problems, to use deterministic models where the mean values interpose only because that generally leads to a very erroneous representation of reality. This chapter presents the basic tools for the development of a model of reliability in biomechanics especially in prosthesis design. This model is considered implicit and requires a numerical simulation to identify the parameters of structural response. To estimate this response, with good accuracy, the finite element method (FEM) appears as a preferential numerical simulation tool. This method consists of discretizing a structure, such as a prosthesis, into a set of subdomains, called finite elements or mesh, linked to each other by nodes. The calculation of a structure is to establish an equation system for the displacements of the set of all meshing nodes and deduct, pursuant to their resolution, the approximations of the deformation and stress fields. Then, the reliability itself is performed by a process of optimization. To perform the optimization and reliability, a sensitivity analysis is required to determine or identify the role of each parameter with respect to the constraints and objective functions.

1.2. Advantages of numerical simulation and optimization

In general, biomechanical models are very complicated. It is necessary to have tools and methods to design these models for analyzing satisfaction levels. With regard to the mechanical behavior of the structure, the engineer or designer has a wide range of methods: methods based on knowledge (empirical laws, databases, etc.), simplified calculation methods (strength of materials), the FEM which is the most widely used and methods of optimization. The implementation and relevant use of the FEM require a certain experience. However, in a highly competitive industrial context, this method allows us to:

– reduce costs (optimization of shapes and material volumes, choice of materials, reduction of the number of prototypes, etc.);

– reduce the time (reduce the iteration number in the design process, directly propose viable solutions from the behavioral perspective, focus testing, etc.);

– improve the quality (ensure the respect of the various functions and constraints in terms of reliability, comfort, ergonomics, etc.).

The scope of the FEM is very large. They have proved their effectiveness in the case of problems simple such as in that of great complexity calculations. This field covers all applications of the structural mechanics (statics, plasticity, composites, dynamics, shock, friction, etc.) and also the mechanics of fluids, rheology, heat exchange, electromagnetic calculations, etc. [POU 99].

The use of the FEM in the medical field has an additional interest. In fact, contrary to the fields of automotive or aeronautics, the designed products are not intended to equip a car or an airplane but a human being. This induced on the first hand that it is often impossible to test a device on humans as we perform a crash test on a car to test the operation of an airbag, for example. On the second hand, for medical devices, in particular the implantable devices, the reliability is essential because maintenance is not possible. For the design of these devices, the FEM brings not only the advantages mentioned previously (cost, time and quality) but also helps provide the anatomical models (part of the body, biological tissues, etc.). The use of such models is common and can simulate the behavior of devices in location [AOU 10, RAM 11].

1.3. Numerical simulation by finite elements

1.3.1. *Use*

The design process causes a succession of choices and decisions that lead to the final definition of the product. Numerical simulation has an important role to play in the realization of these choices and can therefore be used at various stages of this process. Depending on the stage of the design, we can differentiate two types of numerical calculation: the calculation for assistance choice and the calculation for validation [POU 99]. The first consists of comparing technical solutions in order to choose the one that meets the criteria set. Contrary to the calculation for validation, it is not the absolute character of the result which is interesting but rather the comparison of the result with that of one or several competing solutions. This type of use allows us to realize simplified models simulations (geometry, behavior laws of materials, etc.). A complete knowledge of the product being not necessary for the simplified model calculation, it may be undertaken in the early stages of the design and prove its interest. In the case of the calculation for validation, the implementation of a numerical simulation requires as a starting point a semi-complete definition of the studied device (forms, dimensions and materials) and its surroundings (boundary conditions, loads, etc.). It comes at a relatively advanced stage of the project to validate a product definition, and therefore a first set of design choices relative to the specifications. The concept of validation implies a good reliability level of the obtained results. It is therefore appropriate to build a suitable calculation model to reproduce a sufficiently precise mechanical behavior of the piece(s). One step of the model validation is then unavoidable. The use of the FEM for the validation of a product is reflected generally (except some very simple cases) by a fine representation of the geometry and thus a model which is complex and costly in terms of computing resources.

1.3.2. Principle

To treat a problem, FE calculation software requires a certain number of input data. These data consist of a complete description of the mechanical problem to be treated (under a unique formalism to each software) as well as knowledge of the parameters related to the treatment method (for example, the type and distribution of finite elements, the steps of resolution, etc.). The software returns results on the physical quantities of interest in the problem (displacement, temperatures, stresses, etc.). The use of this type of tool therefore requires that the user should be able not only to provide relevant input data, but also to assess the reliability of the obtained results. All that requires a good knowledge of the concerned domain as well as an experience of the theoretical aspects of the method. The problem is considered in the form of a model that first defines the geometry of the structure and the boundary conditions (efforts and displacement). In this case, the FE is more widespread and exploited and it is called the "displacement method". It consists of determining the "displacement field", which means the displacement at each point of the structure. In order to represent the displacement field of a structure, the studied field must be discretized using elements of simple geometric shapes (straight line, triangles, quadrangles, tetrahedra, etc.) and of finite dimensions, called "finite elements". The set of discretization constitutes the "mesh" of the structure. The displacement approximation is then performed independently on each FE. The accuracy of the results of the calculation is directly related, on the one hand to the choice of elements and on the other hand to the quality of the realized mesh (number of elements, distribution in the structure, shape of elements, etc.).

1.3.3. General approach

In general, the development of a FE model can be described by a succession of steps (Figure 1.1). Some of these steps may differ slightly or be reversed depending on the modeling software used [AOU 11]. The first step is to create the model geometry either by drawing or by importing when there are already files of the STEP, IGES, etc. type. Next, it should be to assign material characteristics to each part of the geometry and then to perform the mesh. The mesh step includes the choice of the element type, their number and distribution depending on the geometry. The next step defines the boundary conditions (loads, displacements, fixations, contacts, etc.). It allows us to recreate utilization conditions of the product or the modeled part. When these four steps are performed, the model is constituted and the calculation can be started. In the case where the calculation is not converged (divergence case), it is necessary to return to the previous steps to refine the different parameters: geometry, behavior laws of materials, mesh, or boundary conditions and loadings.

When the calculation is successful, the numerical results (output data) can be analyzed. Following the development of a FE model, a validation step is essential to prove its credibility. It is therefore essential, as a first step, to simulate a known situation to compare the simulation results with proven results (for example, experimental results) and then to make modifications that will harden the model. Only the correlation of these results allows the validation of the model.

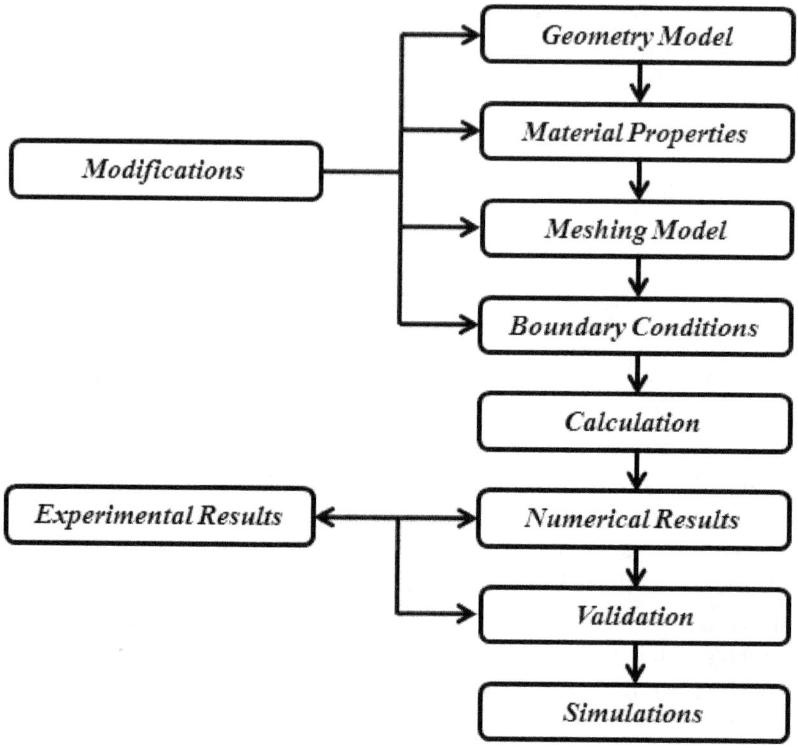

Figure 1.1. *General process for developing a finite element model*

1.4. Optimization process

The history of optimization is as old as humanity. The first record of optimization being performed was by Heron of Alexandria (Greek engineer, mechanic and mathematician from the first century BC).

> "The shortest path that relates a point P to a point Q and that contains a point of a straight line d, is such that at the point of reflection on the straight line d, the incident angle equals to the reflected angle."

Currently, optimization plays a very important role in operational research and can be applied in several fields (aerospace, automotive and marine industries and recently biomechanics). The constant development of the techniques of computer-aided design and optimization strategies fits within this framework. For example, the optimization of structures for more than 50 years raises the greatest interest. Still too little applied to conventional engineering techniques, it is gradually being introduced and so increasing reliability.

1.4.1. *Basic concepts*

In mathematics, an optimization problem is to find an element minimizing or maximizing a given function among a given set of data.

1.4.1.1. *Optimization parameters*

For each optimization problem, we define a set of optimization parameters: objective functions, variables and constraints.

1.4.1.1.1. Objective function

The objective function is a mathematical function of optimization variables. The minimization of this function can be written as follows:

$$\min f(\mathbf{x}) \tag{1.1}$$

It can be a single function ($f(\mathbf{x})$) to minimize, or several functions ($f_1(\mathbf{x})$, $f_2(\mathbf{x})\ldots$).

1.4.1.1.2. Optimization variables

The optimization variables are a set of variables that govern the situation to be modeled. The vector of variables is given by:

$$\mathbf{x} = \{x_1, x_2, \ldots, x_n\} \tag{1.2}$$

The aim is to find the optimal values $\mathbf{x}^* = \{x_1^*, x_2^*, \ldots, x_n^*\}$ which minimize the objective function(s).

1.4.1.1.3. Constraints

The constraints define a domain of feasibility. The points that verify constraints belong to the feasibility domain. There are two types of constraints: equality constraints can be written as follows:

$$h_i(\mathbf{x}) = 0 \Leftrightarrow \begin{Bmatrix} h_1(x_1, x_2, \ldots, x_n) = 0 \\ h_2(x, x_2, \ldots, x_n) = 0 \\ \ldots \\ h_j(x, x_2, \ldots, x_n) = 0 \end{Bmatrix} \quad [1.3]$$

and inequality constraints can be written as follows:

$$g_j(\mathbf{x}) \leq 0 \Leftrightarrow \begin{Bmatrix} g_1(x, x_2, \ldots, x_n) \leq 0 \\ g_2(x, x_2, \ldots, x_n) \leq 0 \\ \ldots \\ g_k(x, x_2, \ldots, x_n) \leq 0 \end{Bmatrix} \quad [1.4]$$

For example, when mechanical stresses are violated in a structure, there may be a case of failure. In this case, the functions of the mechanical stresses may be considered equality constraints, while the displacement constraints can be considered as constraints of inequality to improve the design.

1.4.1.2. *Local or global optimal solutions*

A point x^* of the space $\Re^n$ represents a local minimum, if there exists a neighborhood of x^* denoted by $V(x^*)$ such that:

$$\forall\, x \in V(x^*) \Rightarrow f(x) \geq f(x^*) \quad [1.5]$$

This relation signifies that at the neighborhood of x^*, there exists no point for which $f(x)$ is smaller than $f(x^*)$. A point x^* of the space $\Re^n$ represents a global minimum if:

$$\forall\, x \in \Re^n \Rightarrow f(x) \geq f(x^*) \quad [1.6]$$

There may be several local optimal solutions, but the global optimal solution is unique and often very difficult to find.

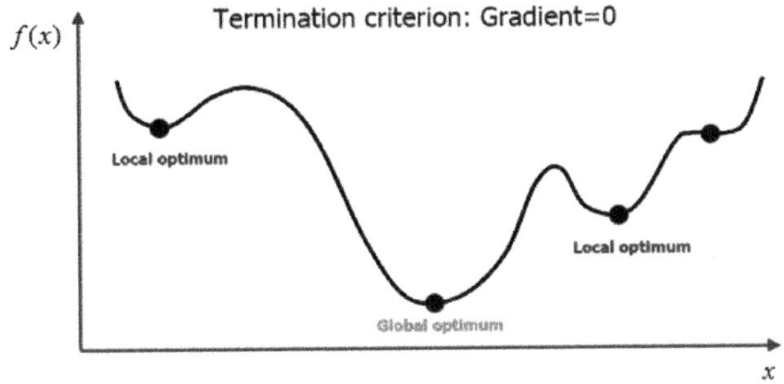

Figure 1.2. *Local optimal solutions and a global optimal solution*

Figure 1.2 shows several local optimal solutions but there exists only one global optimal solution.

1.4.1.3. *Simplified algorithm*

The word "algorithm" comes from the name of the Arabic mathematician "Muhammad Ibn Musa AL-KHAWARIZMI" (9th Century), born in Khwarezm in Uzbekistan.

> "The algorithm is a finite successive series of rules, to apply in a determined order, to a finite number of data values, in order to arrive with certainty, in a finite number of steps, at a certain result independently of the data".

The simplified algorithm of the optimization process can be summarized in five stages:

– the first step represents the data input;

– the second step represents the numerical simulation that yields the results;

– the aim of the third step is to compare the results according to the different convergence conditions;

– the fourth step is to improve the objective function by modifying the different optimization variables;

– the final step is to finish the optimization process and show the optimal results.

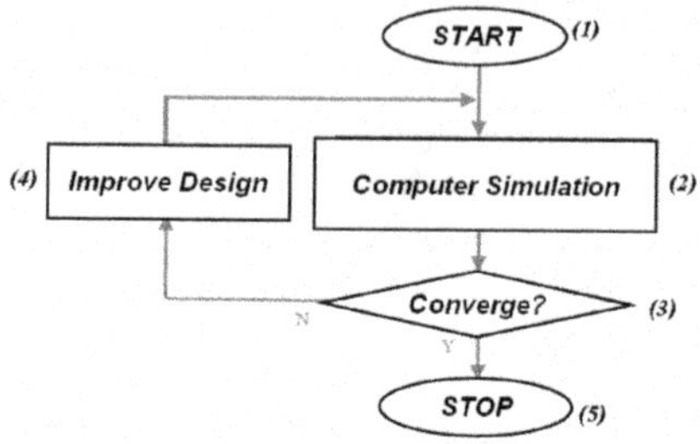

Figure 1.3. *Simplified algorithm of the optimization process*

Figure 1.3 shows the simplified algorithm of an optimization process and its different stages.

1.4.2. *Problem classification*

Optimization problems are classified according to several categories: constraint existence, function linearity (objective and constraints) and objective multiplicity.

1.4.2.1. *Constraint classification*

Optimization problems are classified according to their constraints into two categories: unconstrained and constrained problems.

1.4.2.1.1. Unconstrained problems

Unconstrained optimization has a direct and evident advantage for the identification of peaks (maxima) and troughs (minima), as well as saddle points, before progressing to a more in-depth analysis (peak and trough lines, coarseness, curvatures, geodesics, etc.). The unconstrained optimization problem can be written as follows:

$$\min f(\mathbf{x}) \quad [1.7]$$

The minimization of an objective in the absence of constraints leads to zero values of the components of the first-order derivative vector. If the objective function is nonlinear, a finite solution may exist, even in the absence of constraints. In contrast, the solution is always infinite when the objective function is linear and there are no constraints.

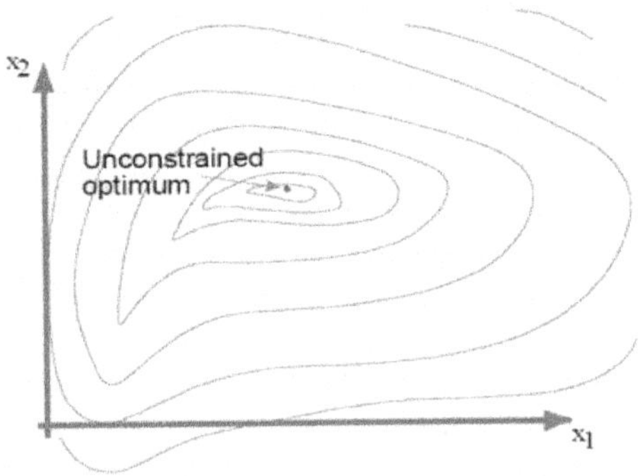

Figure 1.4. *Example of an unconstrained optimization problem*

Figure 1.4 shows an example of an unconstrained optimization problem, with a global optimum.

1.4.2.1.2. Constrained problems

Although a large number of optimization problems present themselves as constrained optimization problems, they can be reduced to unconstrained problems by an increase of variables (Lagrange multipliers for equality constraints and adjustment variables or gap variable for inequality constraints, as discussed in section 1.4.5.2.1). The optimization problem of an objective under constraints, equalities or inequalities can be written using the following form:

$$\begin{array}{ll} \min & : f(\mathbf{x}) \\ \text{subject to} & : g(\mathbf{x}) \le 0 \\ \text{and} & : h(\mathbf{x}) = 0 \end{array} \qquad [1.8]$$

where $h(\mathbf{x}) = 0$ is the equality constraint and $g(\mathbf{x}) \le 0$ is the inequality constraint.

Figure 1.5 shows an example of a constrained optimization problem with an optimum which can be global.

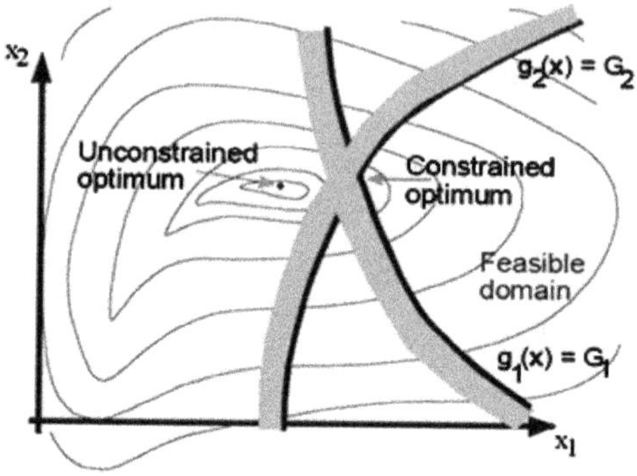

Figure 1.5. *Example of a constrained optimization problem*

ILLUSTRATIVE EXAMPLE 1.1.–

Consider a cylindrical reservoir (or tank) of diameter D and height H (Figure 1.6). The aim is to minimize the reservoir volume subject to the following conditions:

– the diameter must be between 8 and 3.5 cm;

– the height must be between 18 and 8 cm;

– the minimal volume value must not be below 400 ml.

Write the suitable formulation of this optimization problem.

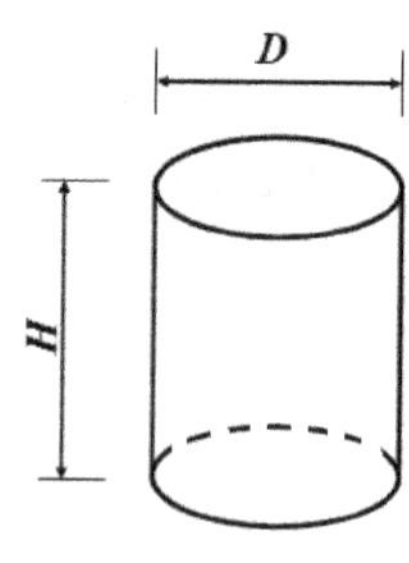

Figure 1.6. *Cylindrical reservoir (or tank)*

SOLUTION.–

The optimization variables are:

$$x_1 = D,\ \ x_2 = H \qquad [1.9]$$

The objective as a function of the variables can be written using the following form:

$$f(D,H) = 2\frac{\pi D^2}{4} + \pi DH,\ \ f(x_1,x_2) = \frac{\pi}{2}x_1^2 + \pi.x_1.x_2 \qquad [1.10]$$

The inequality constraints are:

$$g(D,H) = \frac{\pi D^2}{4}H \geq 400 \qquad [1.11]$$

$$g(x_1,x_2) = \frac{\pi}{4}x_1^2.x_2 \geq 400 \qquad [1.12]$$

$$g(x_1,x_2) = 1 - \frac{\pi}{1600}x_1^2.x_2 \leq 0 \qquad [1.13]$$

The formulation of this constrained optimization problem is mathematically expressed by:

$$\begin{aligned} &\min && :f(x_1,x_2) = \frac{\pi}{2}x_1^2 + \pi.x_1.x_2 \\ &\text{subject to} && :g(x_1,x_2) = 1 - \frac{\pi}{1600}x_1^2.x_2 \leq 0 \\ & && :3 \leq x_1 \leq 8 \\ & && :8 \leq x_1 \leq 18 \end{aligned} \qquad [1.14]$$

1.4.2.2. *Linearity classification*

Optimization problems are classified according to the function linearity (objective and constraints): linear optimization or nonlinear optimization.

1.4.2.2.1. Linear problems

A function $f(x_1, x_2, ..., x_n)$ of $x_1, x_2, ..., x_n$ is a linear function if and only if there exists a set of constants $c_1, c_2, ..., c_n$ such that:

$$f(x_1, x_2, ..., x_n) = c_1 x_1 + c_2 x_2 ... + c_n x_n \qquad [1.15]$$

In the linear problem, all functions (objective, equality and inequality constraints) must be linear. For example, consider a linear problem expressed by the following form:

$$\begin{aligned} &\min : f(x) = -3x_1 - x_2 \\ &\text{s.t. } : \begin{cases} x_2 \leq 9 \\ x_1 = x_2 \end{cases} \end{aligned} \qquad [1.16]$$

In this problem, all functions (objective and equality and inequality constraints) are linear (first order).

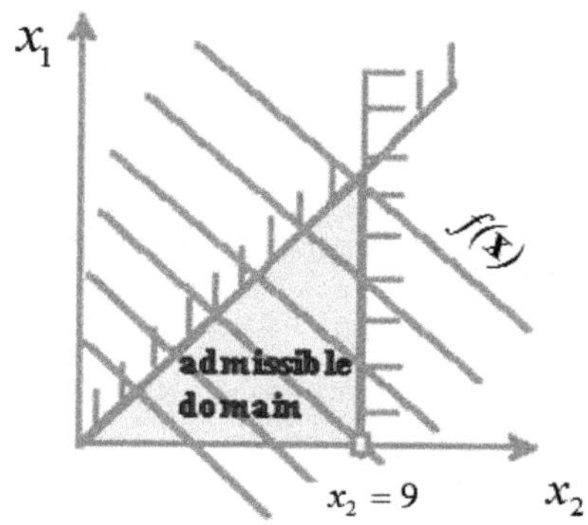

Figure 1.7. *Example of a linear problem*

Figure 1.7 shows an example of a linear problem where all functions are modeled by straight lines.

1.4.2.2.2. Nonlinear problems

In a nonlinear problem, one or more functions (objective, equality and inequality constraints) must be nonlinear. For example, consider a nonlinear problem expressed by the following form:

$$\begin{aligned} &\min: f(x) = -3x_1 - x_2 \\ &\text{s.t.} \ : \begin{cases} x_2 \le 9 \\ x_1^2 \le x_2 \end{cases} \end{aligned} \qquad [1.17]$$

In this problem, we only have one function that is nonlinear (second order). Although the other functions are linear, we consider this problem to be nonlinear.

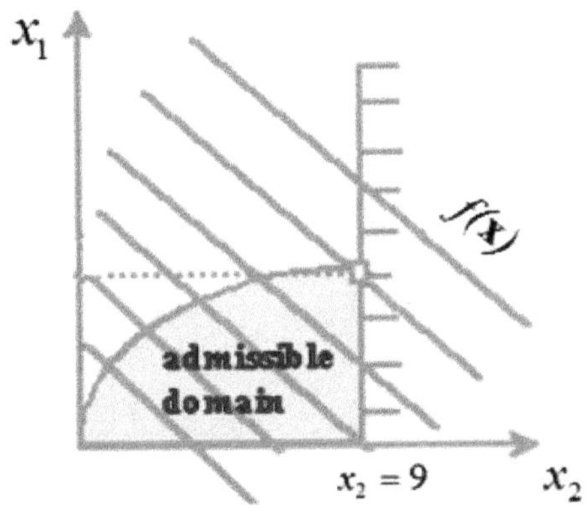

Figure 1.8. *Example of a nonlinear problem*

Figure 1.8 shows an example of a nonlinear problem where one of the functions is modeled by a second-order curve.

1.4.2.3. *Objective classification*

Optimization problems can also be classified according to the multiplicity of the objectives: single-objective or multi-objective optimization.

1.4.2.3.1. Single-objective problems

We can easily solve a problem with only one objective and identify the optimal solutions. In this case, the objective is represented by a single component:

$$f(\mathbf{x}) \qquad [1.18]$$

Whatever the number of associated constraints, we consider this problem to be single-objective.

1.4.2.3.2. Multi-objective problems

We have several criteria when solving multi-objective problems. In this case, the objective is represented by a vector with multiple components:

$$\begin{Bmatrix} f_1(\mathbf{x}) \\ f_2(\mathbf{x}) \\ \vdots \\ \vdots \\ f_z(\mathbf{x}) \end{Bmatrix} \qquad [1.19]$$

This type of problem is often encountered in biomechanics.

ILLUSTRATIVE EXAMPLE 1.2.–

Consider an optimization problem in the following form:

$$\begin{aligned} &\max_{x,y} \;\; 20x+9y \\ &\text{subject to:} \\ &\quad 3x+y \leq 20 \\ &\quad 2x+y \leq 15 \\ &\quad\quad x,y \geq 0 \end{aligned} \qquad [1.20]$$

– classify this optimization problem;

– model geometrically all functions (objective and constraints);

– find the optimal solution.

SOLUTION.–

The classification of this optimization problem is a constrained, linear and single-objective problem.

To model this problem geometrically, we model all functions in the Cartesian space. We start with the function of the first constraint:

$$3x + y \leq 20 \quad [1.21]$$

The intersection of the equation $3x + y = 20$ with the Cartesian axes divides space into two domains: a feasible domain and an infeasible domain (Figure 1.9). To test the two domains, we choose the origin coordinates $(0,0)$.

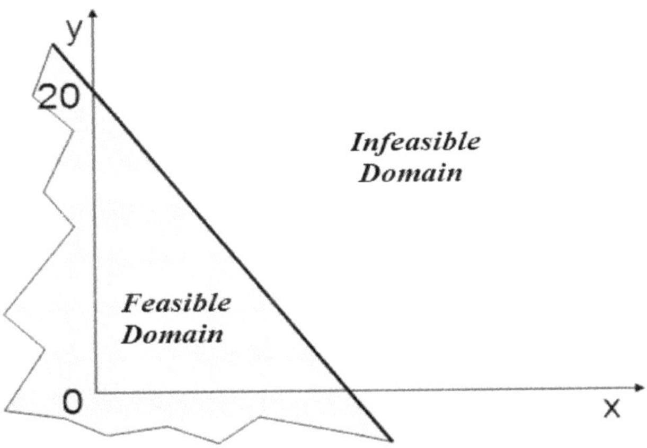

Figure 1.9. *Model of the first constraint*

Next, we model the function of the second constraint:

$$2x + y \leq 15 \quad [1.22]$$

The intersection of the equation $2x + y = 15$ with the Cartesian axes divides space into two domains: a feasible domain and an infeasible domain.

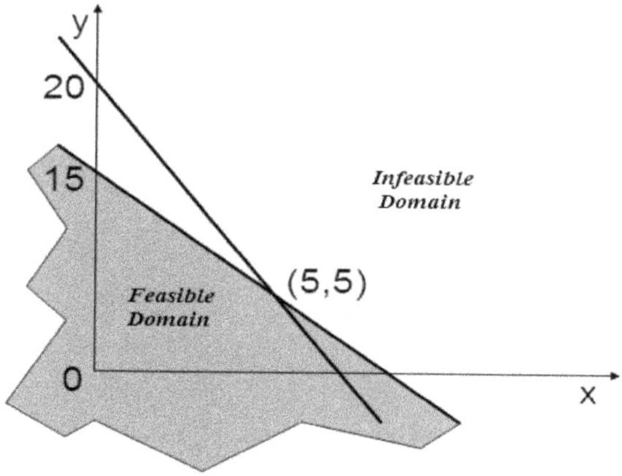

Figure 1.10. *Model of the second constraint*

Afterward, we find the intersection of the two constraints:

$$\begin{aligned} 3x + y &\leq 20 \\ 2x + y &\leq 15 \end{aligned} \quad [1.23]$$

Figure 1.11 shows the intersection of two straight line segments at the point (5,5).

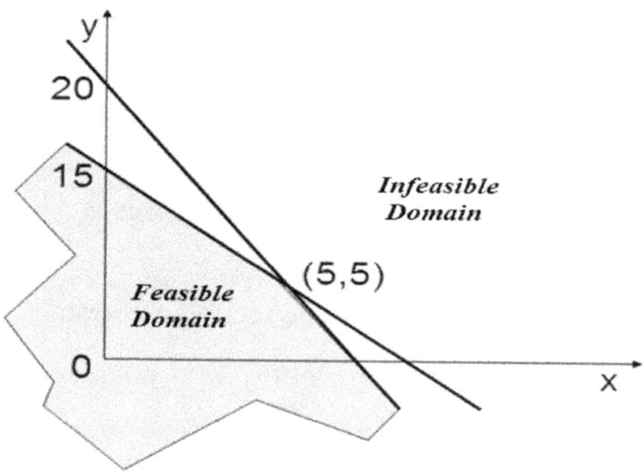

Figure 1.11. *Model of the intersection of two constraints*

Next, we model all constraints to find the feasible domain:

$$\begin{aligned} 3x + y &\leq 20 \\ 2x + y &\leq 15 \\ x, y &\geq 0 \end{aligned} \qquad [1.24]$$

Figure 1.12 shows the intersection of the set of constraints.

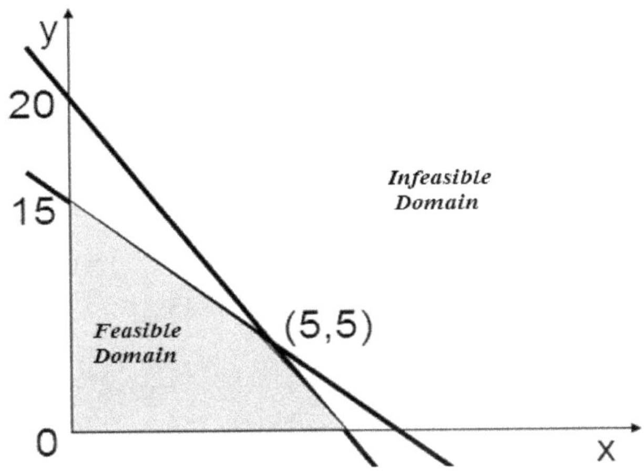

Figure 1.12. *Model of the intersection of the set of constraints*

Finally, we model the whole problem:

$$\begin{aligned} &\max_{x,y} \; 20x + 9y \\ &\text{subject to:} \\ &\quad 3x + y \leq 20 \\ &\quad 2x + y \leq 15 \\ &\quad x, y \geq 0 \end{aligned} \qquad [1.25]$$

Figure 1.13 shows the model of the iso-values of the objective function ($f(x,y) = 20x + 9y$) with the intersection of the set of constraints.

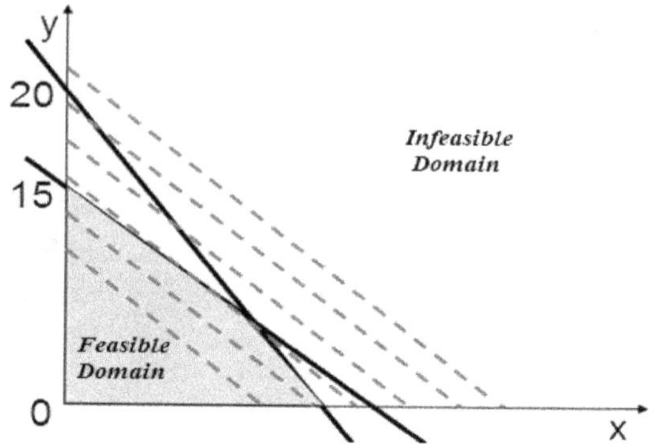

Figure 1.13. *Model of the iso-values of the objective function with the set of constraints*

Figure 1.14 shows the optimal solution ($(x^*, y^*) = (5,5)$) where the maximal value of the objective is $f(5,5) = 145$.

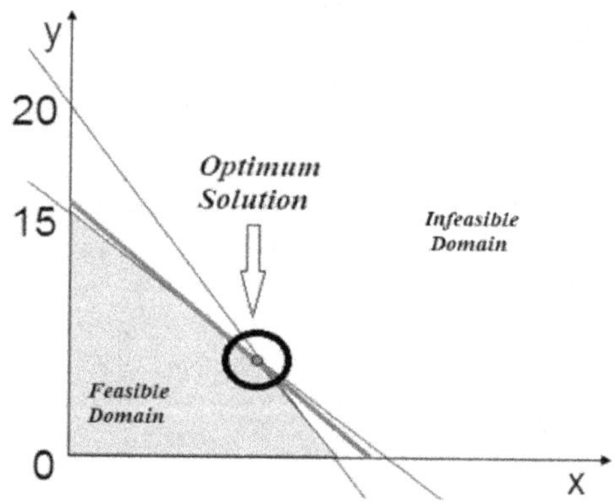

Figure 1.14. *Model of the optimal point*

ILLUSTRATIVE EXAMPLE 1.3.–

Consider the optimization problem expressed by the following form:

$$\min f(\mathbf{x}) = e^{-x} + x^2 \quad [1.26]$$

– classify this optimization problem;

– model geometrically the objective function $f(\mathbf{x})$;

– find the optimal solution.

SOLUTION.–

The classification of this optimization problem is an unconstrained, nonlinear, single-objective problem.

To model this function geometrically, we calculate the function for several values of the variable x (Table 1.1).

x	0	0.5	1	2	3
$f(x)$	1	0.83	1.36	4.13	9.03

Table 1.1. *Objective function evaluation for different variable values*

We observe that the objective function changes its direction when considering the different values of the variable x. In order to determine the minimum value of the objective function, we can analytically calculate its derivative as follows:

$$\frac{\partial f}{\partial x} = -e^{-x} + 2x \qquad [1.27]$$

Considering the null value of the derivative, we obtain:

$$x = \frac{1}{2}e^{-x} \qquad [1.28]$$

This equation can be solved using iterative techniques (Table 1.2).

Iteration	0	1	2	3	4	5	6	7	8
x	0	0.5	0.303	0.369	0.349	0.354	0.350	0.351	0.351

Table 1.2. *Iterative values until convergence*

The optimal value of the function is $f(x^*) = 0.828$ when $x^{(8)} = 0.351$. Figure 1.15 shows the geometrical interpolation of the objective function and the optimal value of the variable x.

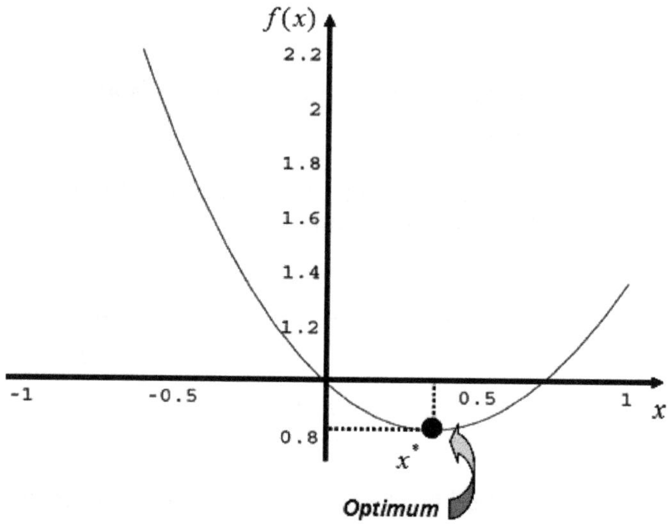

Figure 1.15. Objective *function interpretation and the optimum value*

1.4.3. *Optimization methods*

We select an optimization method according to the following four criteria:

– the *generality* is the potential to use the method in different applications;

– the *robustness* is the capacity of the method to find optimal solutions regardless of the starting points;

– the *efficiency* is the speed of the method to find optimal solutions;

– the *capacity* is the potential to deal with a large number of variables and constraints.

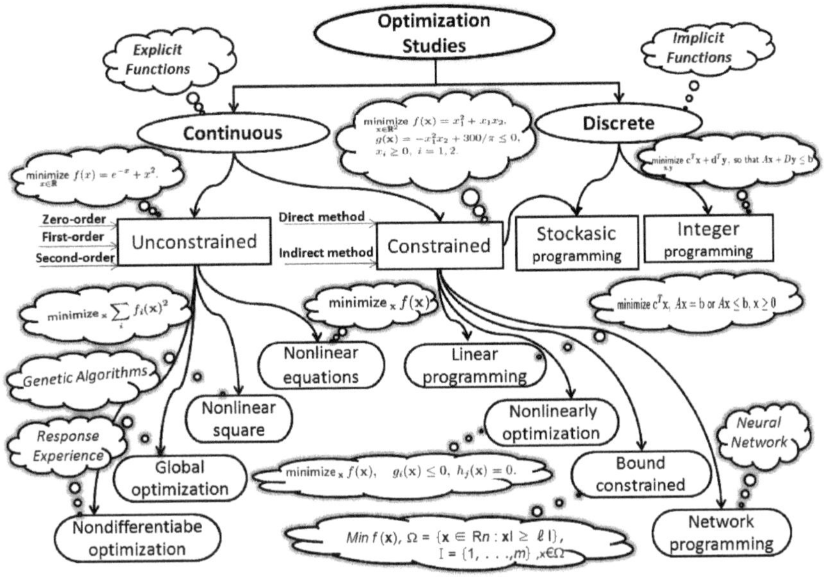

Figure 1.16. *Optimization strategies*

Figure 1.16 presents the different aspects of the optimization process to be studied. In general, we divide the optimization studied into two categories: continuous and discrete optimization [KHA 11]. The continuous optimization needs several analytical competences to simplify the studied problem before solving it, while the discrete optimization necessitates a good computing technology to solve the studied problem.

1.4.4. *Unconstrained methods*

The majority of optimization problems are solved by this type of method. This is partly because constrained optimization problems can be reduced to unconstrained problems. The main advantage of these methods is their potential to be implemented on a machine by a single loop. In this chapter, we only present three categories of these methods.

1.4.4.1. *Zero-order methods*

In this type of method, we use iterative evaluations of functions. We never use the derivatives of functions. These methods can be used for discontinuous and/or non-differentiable functions.

1.4.4.1.1. Simplex method

This method is principally used to solve linear problems. The idea is to calculate three points and then find the direction of the optimal solution as is shown in Figure 1.17.

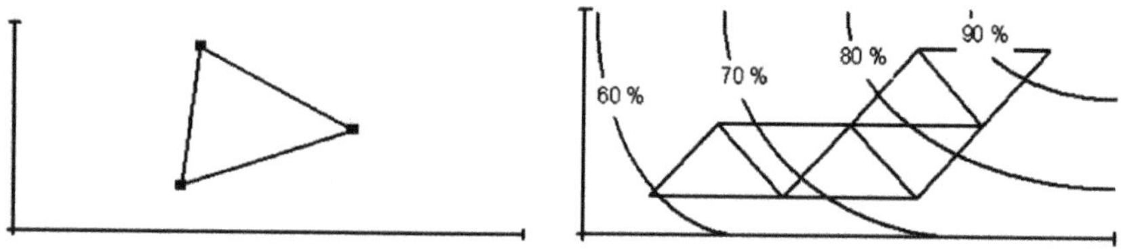

Figure 1.17. *Principle of the simplex method*

1.4.4.1.2. Curve fitting method

The principle of this method is based upon curve approximation techniques, which are different relative to the interpolation techniques. The main difference is that the resulting smooth curves should not pass through the used experiment points.

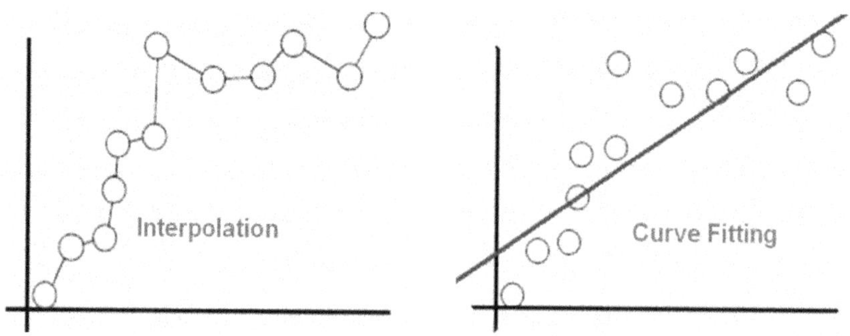

Figure 1.18. *Interpolation and the curve fitting approximation*

Figure 1.18 shows the difference between the approximation and interpolation methods.

1.4.4.2. *First-order methods*

The principle of this type of method is to use the first-order derivative and a progress (forward) step to arrive to the optimal solution. The main algorithm consists of:

– initializing the start point: $k = 0$ and $\mathbf{x}_0$;

– solving for the research direction: $\mathbf{s}_0$;

– evaluating the subsequent point using the equation: $\mathbf{x}_{k+1} = \mathbf{x}_k - \alpha_{k+1}\mathbf{s}_k$;

– passing to the following iteration: $k = k + 1$;

– *while* $\left|x_k - x_{k-1}\right| > \varepsilon$.

1.4.4.2.1. Descent gradient method

The descent gradient method (DGM) is based upon the first-order derivative and a progress step. The principal equation is expressed by the following:

$$\mathbf{x}_{k+1} = \mathbf{x}_k - a_{k+1}\mathbf{s}_k \qquad [1.29]$$

with the iteration $k = 0,1,2,\ldots\ldots,n$ and the search direction is expressed using the following form:

$$\mathbf{d}_{k+1} = \mathbf{x}_{k+1} - \mathbf{x}_k = -a_{k+1}\mathbf{s}_k \qquad [1.30]$$

Here, the search direction is evaluated using the first-order derivative as follows:

$$\mathbf{s}_k = \nabla \mathbf{f}\,(\mathbf{x}_k) \qquad [1.31]$$

and the step size can be calculated by:

$$a_{k+1} = \frac{df(\,a_{k+1})}{d\,a_{k+1}}, \quad k = 0,1,2,\ldots\ldots,n \qquad [1.32]$$

This method has a slow progress and requires a significant computation time.

1.4.4.2.2. Conjugate gradient method

The principle of the conjugate gradient method (CGM) is based upon the evaluation of the first order for the first step ($k = 0$):

$$\mathbf{s}_k = \nabla \mathbf{f}\,(\mathbf{x}_k) \tag{1.33}$$

However, use:

$$\mathbf{s}_{k+1} = \nabla \mathbf{f}\,(\mathbf{x}_{k+1}) + \beta_{k+1}\mathbf{s}_k \tag{1.34}$$

where the coefficient β can be calculated by:

$$\beta = \frac{\langle \nabla f\,(\mathbf{x}_{k+1})\rangle\{\nabla f\,(\mathbf{x}_{k+1})\}}{\langle \nabla f\,(\mathbf{x}_{k})\rangle\{\nabla f\,(\mathbf{x}_{k})\}} = \frac{\|\nabla f\,(\mathbf{x}_{k+1})\|}{\|\nabla f\,(\mathbf{x}_{k})\|} \tag{1.35}$$

In this case, we can use two derivatives in two successive steps. Figure 1.19 shows a geometric interpretation of an example of two variables when considering the DGM and CGM.

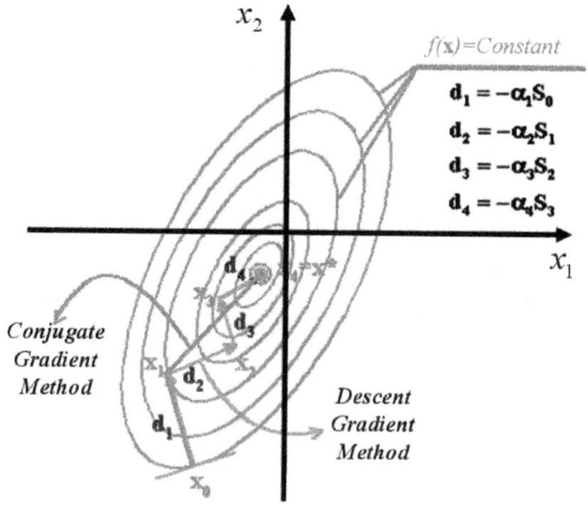

Figure 1.19. *Comparison of the descent gradient and conjugated gradient methods*

For the same starting point $\mathbf{x}_0$, the DGM requires four iterations to arrive at the optimal point $\mathbf{x}^*$, while the CGM only needs two iterations.

1.4.4.3. *Second-order methods*

This type of method requires the evaluation of second-order derivatives. We use a matrix known as the Hessian Matrix as follows:

$$\mathbf{H}(\mathbf{x}_0) = \nabla^2 \mathbf{f}(\mathbf{x}_0) = \begin{Bmatrix} \frac{\partial 2f}{\partial x_1^2}(\mathbf{x}_0) & \frac{\partial 2f}{\partial x_1 \partial x_2}(x_0) & \cdots & \cdots & \frac{\partial 2f}{\partial x_1 \partial x_n}(\mathbf{x}_0) \\ \frac{\partial 2f}{\partial x_1 \partial x_2}(\mathbf{x}_0) & \ddots & & & \vdots \\ \vdots & & \ddots & & \vdots \\ \vdots & & & \ddots & \vdots \\ \frac{\partial 2f}{\partial x_1 \partial x_n}(\mathbf{x}_0) & \cdots & \cdots & \cdots & \frac{\partial 2f}{\partial x_n^2}(\mathbf{x}_0) \end{Bmatrix} \quad [1.36]$$

Despite this type of method having a good level of efficiency, it is considered very sensitive and can lead to instabilities.

1.4.4.3.1. Newton's method

Isaac Newton (1643–1727) was an English physicist, astronomer and mathematician. He is often considered to be the most influential scientist of all time. His contributions include differential calculus, algebra, analytic geometry, his law of gravitation and laws of planetary motion.

Newton's method (NM) is also based upon the basic iterative equation as follows:

$$\mathbf{x}_{k+1} = \mathbf{x}_k - a_{k+1}\mathbf{s}_k \quad [1.37]$$

However, the search direction depends on the Hessian matrix as follows:

$$\{s_k\} = [H(\mathbf{x}_k)]^{-1} \{\nabla f(\mathbf{x}_k)\} \quad [1.38]$$

We consider the following forward step:

$$a_{k+1} = 1 \quad [1.39]$$

Hence, the iterative equation can be written as follows:

$$\mathbf{x}_{k+1} = \mathbf{x}_k - [H(x_k)]^{-1}\{\nabla f(\mathbf{x}_k)\} \qquad [1.40]$$

NM is efficient but suffers from a repeated calculation of the Hessian matrix at each iteration.

1.4.4.3.2. Modified Newton's method

The modified Newton's method (MNM) requires a single calculation of the Hessian matrix at the first iteration. The iterative equation is written using the following form:

$$\mathbf{x}_{k+1} = \mathbf{x}_k - \alpha_{k+1}[H(\mathbf{x}_0)]^{-1}\{\nabla f(\mathbf{x}_k)\} \qquad [1.41]$$

with a search direction:

$$\mathbf{s}_k = [H(\mathbf{x}_0)]^{-1}\{\nabla f(\mathbf{x}_k)\} \qquad [1.42]$$

If $\mathbf{s}_k = \{\nabla f(\mathbf{x}_k)\}$, we return to the DGM, and if $\alpha_{k+1} = 1$ and $\mathbf{s}_k = [H(\mathbf{x}_k)]^{-1}\{\nabla f(\mathbf{x}_k)\}$, we return to NM.

ILLUSTRATIVE EXAMPLE 1.4.–

Consider an unconstrained optimization problem using the following form:

$$\min : f(\mathbf{x}) = x_1 - x_2 + 2x_1^2 + 2x_1x_2 + x_2^2 \qquad [1.43]$$

– find the minimal value using first- and second-order methods using the starting point $\mathbf{x}_0 = (0,0)$;

– compare the different methods according to computing time.

SOLUTION.–

We start with the first-order methods (DGM and CGM), then the second-order methods (NM and MNM).

Descent gradient method

We first find the first-order derivative of the objective function $f(\mathbf{x})$:

$$\nabla f(\mathbf{x}) = \begin{Bmatrix} \dfrac{\partial f(\mathbf{x})}{\partial x_1} \\ \dfrac{\partial f(\mathbf{x})}{\partial x_2} \end{Bmatrix} = \begin{Bmatrix} 1+4x_1+2x_2 \\ -1+2x_1+2x_2 \end{Bmatrix} \quad [1.44]$$

Next, we calculate the search direction for the initial point:

$$S_0 = \nabla f(\mathbf{x_0}) = \begin{Bmatrix} 1 \\ -1 \end{Bmatrix} \quad [1.45]$$

By substituting this result into the iterative equation, we obtain:

$$x_{(1)} = x_{(0)} - \alpha_1 S_0 \quad [1.46]$$

When considering the forward step as a variable, we obtain:

$$x_{(1)} = \begin{Bmatrix} 0 \\ 0 \end{Bmatrix} - \alpha_1 \begin{Bmatrix} 1 \\ -1 \end{Bmatrix} = \begin{Bmatrix} -\alpha_1 \\ \alpha_1 \end{Bmatrix} \quad [1.47]$$

We calculate the forward step by substituting the optimization variables in the objective function formulation as follows:

$$f(\alpha_1) = -\alpha_1 - \alpha_1 + 2\alpha_1^2 - 2\alpha_1^2 + \alpha_1^2 = +\alpha_1^2 - 2\alpha_1 \quad [1.48]$$

The derivative of the iterative equation with respect to the current step gives:

$$\frac{\partial f(\alpha_1)}{\partial \alpha_1} = 2\alpha_1 - 2 = 0 \Rightarrow \alpha_1 = 1 \quad [1.49]$$

The new value is:

$$x_{(1)} = \begin{Bmatrix} -1 \\ 1 \end{Bmatrix} \quad [1.50]$$

Figure 1.20 shows the coordinates of the new value at iteration (1).

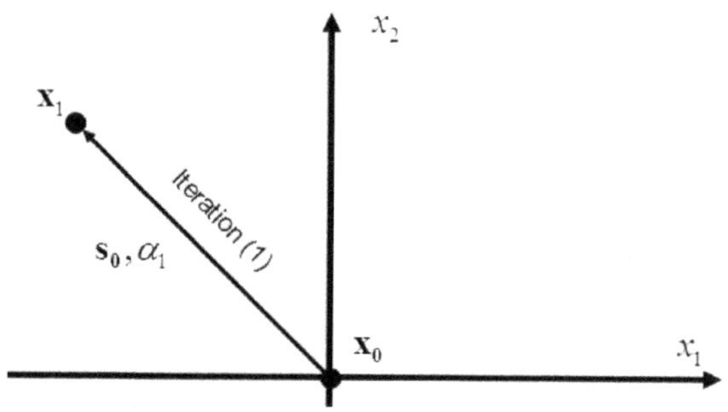

Figure 1.20. *Iteration (1) using DGM*

Next, we calculate the search direction for iteration 1:

$$S_1 = \nabla f(\mathbf{x}_1) = \begin{Bmatrix} -1 \\ -1 \end{Bmatrix} \quad [1.51]$$

By substituting this result into the iterative equation, we obtain:

$$x_{(2)} = x_{(1)} - \alpha_2 S_{(1)} \quad [1.52]$$

When considering the forward step as a variable, we obtain:

$$x_{(2)} = \begin{Bmatrix} -1 \\ 1 \end{Bmatrix} - \alpha_2 \begin{Bmatrix} -1 \\ -1 \end{Bmatrix} = \begin{Bmatrix} -1+\alpha_2 \\ +1+\alpha_2 \end{Bmatrix} \quad [1.53]$$

We calculate the forward step by substituting the optimization variables in the objective function formulation as follows:

$$f(\alpha_2) = (-1+\alpha_2) - (1+\alpha_2) + 2(-1+\alpha_2)^2 + 2(-1+\alpha_2)(1+\alpha_2) + (1+\alpha_2)^2 \quad [1.54]$$

The derivative of the iterative equation with respect to the current step gives:

$$\frac{\partial f(\alpha_2)}{\partial \alpha_2} = -10\alpha_2 - 2 = 0 \Rightarrow \alpha_2 = \frac{1}{5} \quad [1.55]$$

The new value is:

$$x_{(2)} = \begin{Bmatrix} -0.8 \\ 1.2 \end{Bmatrix} \quad [1.56]$$

Figure 1.21 shows the coordinates of the new value at iteration (2).

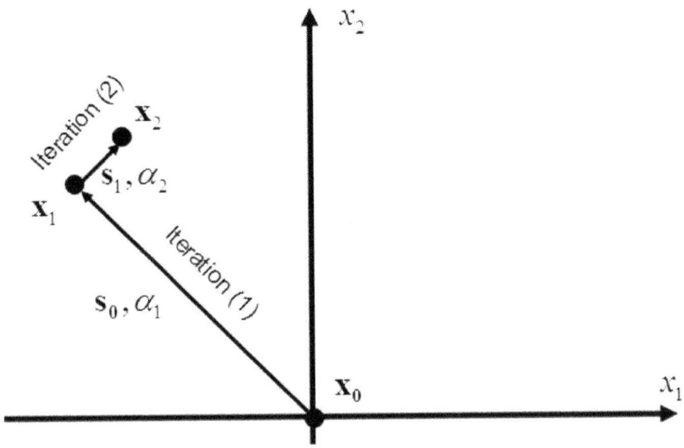

Figure 1.21. *Iteration (2) using DGM*

Next, we calculate the search direction for iteration 2:

$$S_{(2)} = \nabla f(\mathbf{x}_2) = \begin{Bmatrix} 0.2 \\ -0.2 \end{Bmatrix} \quad [1.57]$$

By substituting this result into the iterative equation, we obtain:

$$x_{(3)} = x_{(2)} - \alpha_3 S_{(2)} \quad [1.58]$$

When considering the forward step as a variable, we obtain:

$$x_{(3)} = \begin{Bmatrix} -0.8 \\ 1.2 \end{Bmatrix} - \alpha_3 \begin{Bmatrix} 0.2 \\ -0.2 \end{Bmatrix} = \begin{Bmatrix} -0.8 - 0.2\alpha_3 \\ 1.2 + 0.2\alpha_3 \end{Bmatrix} \quad [1.59]$$

We calculate the forward step by substituting the optimization variables in the objective function formulation as follows:

$$f(\alpha_3)=(-0.8-0.2\alpha_3)-(1.2+0.2\alpha_3)+2(-0.8-0.2\alpha_3)^2 \\ +2(-0.8-0.2\alpha_3)(1.2+0.2\alpha_3)+2(1.2+0.2\alpha_3)^2 \quad [1.60]$$

The obtained equation can be simplified into the following form:

$$f(\alpha_3)=0.24+0.4\alpha_3+0.08\alpha_3^2 \quad [1.61]$$

The derivative of the iterative equation with respect to the current step gives:

$$\frac{\partial f(\alpha_3)}{\partial \alpha_3}=0.4+0.16\alpha_3=0 \Rightarrow \alpha_3=-2.5 \quad [1.62]$$

The new value is:

$$\mathbf{x}_{(3)}=\begin{Bmatrix}-0.3\\0.7\end{Bmatrix} \quad [1.63]$$

Figure 1.22 shows the coordinates of the new value at iteration (3).

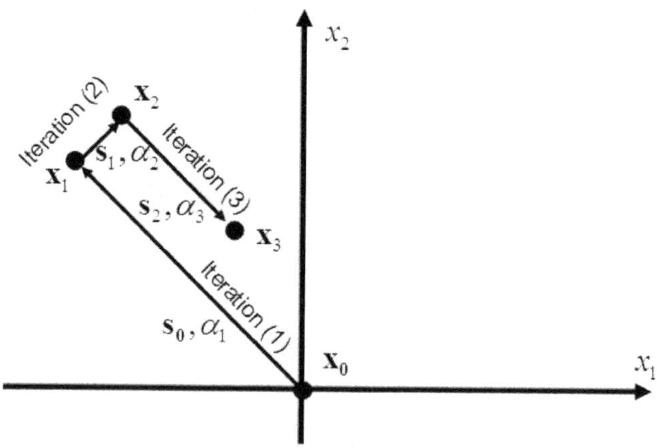

Figure 1.22. *Iteration (3) using DGM*

Next, we calculate the search direction for iteration 3:

$$S_{(3)} = \nabla f(\mathbf{x}_3) = \begin{Bmatrix} 1.2 \\ -0.2 \end{Bmatrix} \qquad [1.64]$$

By substituting this result into the iterative equation, we obtain:

$$x_{(4)} = x_{(3)} - \alpha_4 S_{(3)} \qquad [1.65]$$

When considering the forward step as a variable, we obtain:

$$x_{(4)} = \begin{Bmatrix} -0.3 \\ 0.7 \end{Bmatrix} - \alpha_4 \begin{Bmatrix} 1.2 \\ -0.2 \end{Bmatrix} = \begin{Bmatrix} -0.3 - 1.2\alpha_4 \\ 0.7 + 0.2\alpha_4 \end{Bmatrix} \qquad [1.66]$$

Figure 1.23 shows the coordinates of the new value at iteration (4).

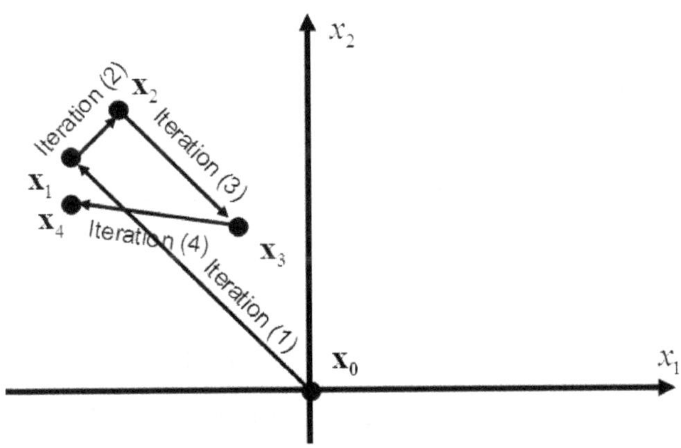

Figure 1.23. *Iteration (4) using DGM*

We calculate the forward step by substituting the optimization variables in the objective function formulation as follows:

$$\begin{aligned} f(\alpha_4) = & (-0.3 - 1.2\alpha_4) \; - (0.7 + 0.2\alpha_4) + 2(-0.3 - 1.2\alpha_4)^2 \\ & + 2(-0.3 - 1.2\alpha_4)(0.7 + 0.2\alpha_4) + 2(0.7 + 0.2\alpha_4)^2 \end{aligned} \qquad [1.67]$$

The obtained equation can be simplified into the following form:

$$f(\alpha_4)=1.04\alpha_4^2-1.2\alpha_4-0.26 \quad [1.68]$$

The derivative of the iterative equation with respect to the current step gives:

$$\frac{\partial f(\alpha_4)}{\partial \alpha_4}=2.08\alpha_4-1.2=0 \Rightarrow \alpha_4=0.58 \quad [1.69]$$

The new value is:

$$\mathbf{x}_{(4)}=\begin{Bmatrix}-0.996\\ 0.816\end{Bmatrix} \quad [1.70]$$

Next, we calculate the search direction for iteration 4:

$$S_{(4)}=\nabla f(\mathbf{x}_4)=\begin{Bmatrix}-1.352\\ -1.36\end{Bmatrix} \quad [1.71]$$

By substituting this result into the iterative equation, we obtain:

$$x_{(5)}=x_{(4)}-\alpha_5 S_4 \quad [1.72]$$

When considering the forward step as a variable, we obtain:

$$x_{(5)}=\begin{Bmatrix}-0.996\\ 0.816\end{Bmatrix}-\alpha_5\begin{Bmatrix}-1.352\\ -1.36\end{Bmatrix}=\begin{Bmatrix}-0.996+1.352\alpha_5\\ 0.816+1.36\alpha_5\end{Bmatrix} \quad [1.73]$$

We calculate the forward step by substituting the optimization variables in the objective function formulation as follows:

$$\begin{aligned}f(\alpha_5)=&(-0.996+1.352\alpha_5)-(0.816+1.36\alpha_5)+2(-0.996+1.352\alpha_5)^2\\&+2(-0.996+1.352\alpha_5)(0.816+1.36\alpha_5)+2(0.816+1.36)^2\end{aligned} \quad [1.74]$$

The equation obtained can be simplified into the following form:

$$f(\alpha_5) = -0.122 - 3.078\alpha_5 + 11.02\alpha_5^2 \qquad [1.75]$$

The derivative of the iterative equation with respect to the current step gives:

$$\frac{\partial f(\alpha_5)}{\partial \alpha_5} = 22.04\alpha_5 - 3.078 = 0 \Rightarrow \alpha_5 = 0.14 \qquad [1.76]$$

The new value is:

$$\mathbf{x}_{(5)} = \begin{Bmatrix} -0.8 \\ 1 \end{Bmatrix} \qquad [1.77]$$

Figure 1.24 shows the coordinates of the new value at iteration (5).

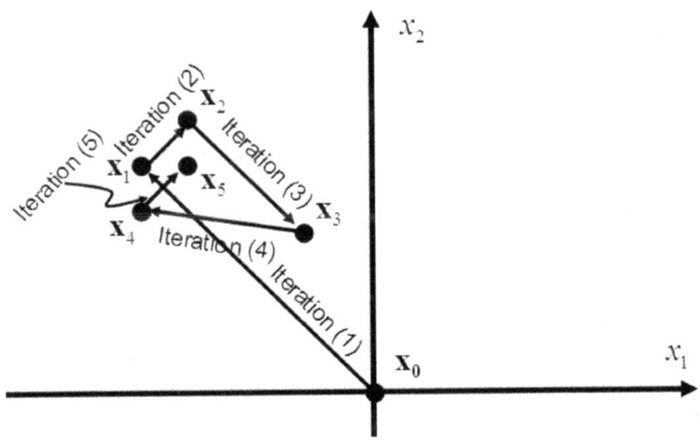

Figure 1.24. *Iteration (5) using DGM*

Next, we calculate the search direction for iteration 5:

$$S_{(5)} = \nabla f(\mathbf{x}_5) = \begin{Bmatrix} -0.2 \\ -0.6 \end{Bmatrix} \qquad [1.78]$$

By substituting this result into the iterative equation, we obtain:

$$x_{(6)} = x_{(5)} - \alpha_6 S_5 \qquad [1.79]$$

When considering the forward step as a variable, we obtain:

$$x_{(6)} = \begin{Bmatrix} -0.8 \\ 1 \end{Bmatrix} - \alpha_6 \begin{Bmatrix} -0.2 \\ -0.6 \end{Bmatrix} = \begin{Bmatrix} -0.8 + 0.2\alpha_6 \\ 1 + 0.6\alpha_6 \end{Bmatrix} \qquad [1.80]$$

We calculate the forward step by substituting the optimization variables in the objective function formulation as follows:

$$\begin{aligned} f(\alpha_6) = {} & (-0.8 + 0.2\alpha_6) \; - (1 + 0.6\alpha_6) + 2(-0.8 + 0.2\alpha_6)^2 \\ & + 2(-0.8 + 0.2\alpha_6)(1 + 0.6\alpha_6) + (1 + 0.6\alpha_6)^2 \end{aligned} \qquad [1.81]$$

The obtained equation can be simplified into the following form:

$$f(\alpha_6) = 1.04\alpha_6^2 + 0.8\alpha_6 - 0.12 \qquad [1.82]$$

The derivative of the iterative equation with respect to the current step gives:

$$\frac{\partial f(\alpha_6)}{\partial \alpha_6} = 1.04\alpha_6 + 0.8 = 0 \Rightarrow \alpha_6 = 0.77 \qquad [1.83]$$

The new value is:

$$\mathbf{x}_{(6)} = \begin{Bmatrix} -1 \\ 1.5 \end{Bmatrix} \qquad [1.84]$$

At the optimal point, we find that the derivative is null:

$$\nabla f(\mathbf{x}_6) = \begin{Bmatrix} 0 \\ 0 \end{Bmatrix} \qquad [1.85]$$

Figure 1.25 shows the coordinates of the new value at iteration (6).

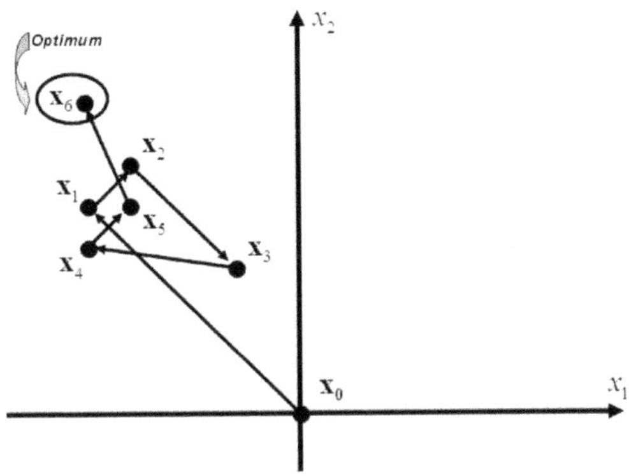

Figure 1.25. *Iteration (6) using DGM*

Conjugate gradient method

We calculate the search direction for the initial point considering the derivative as follows:

$$S_0 = \nabla f(\mathbf{x_0}) = \begin{Bmatrix} 1 \\ -1 \end{Bmatrix} \quad [1.86]$$

When considering the forward step as a variable, we obtain:

$$x_{(1)} = \begin{Bmatrix} 0 \\ 0 \end{Bmatrix} - \alpha_1 \begin{Bmatrix} 1 \\ -1 \end{Bmatrix} = \begin{Bmatrix} -\alpha_1 \\ \alpha_1 \end{Bmatrix} \quad [1.87]$$

We calculate the forward step by substituting the optimization variables in the objective function formulation as follows:

$$f(\alpha_1) = -\alpha_1 - \alpha_1 + 2\alpha_1^2 - 2\alpha_1^2 + \alpha_1^2 = +\alpha_1^2 - 2\alpha_1 \quad [1.88]$$

The derivative of the iterative equation with respect to the current step gives:

$$\frac{\partial f(\alpha_1)}{\partial \alpha_1} = 2\alpha_1 - 2 = 0 \Rightarrow \alpha_1 = 1 \quad [1.89]$$

The new value is:

$$x_{(1)} = \begin{Bmatrix} -1 \\ 1 \end{Bmatrix} \qquad [1.90]$$

The second stage deals with calculating the coefficient β. Here, we determine the derivative at this point:

$$S_1 = \nabla f(\mathbf{x}_1) = \begin{Bmatrix} -1 \\ -1 \end{Bmatrix} \qquad [1.91]$$

Next, we calculate the coefficient β using the following equation:

$$\beta_{k+1} = \frac{\langle \nabla f(x_k) \rangle \{\nabla f(x_k)\}}{\langle \nabla f(x_{k-1}) \rangle \{\nabla f(x_{k-1})\}} = \frac{\langle -1 \quad -1 \rangle \begin{Bmatrix} -1 \\ -1 \end{Bmatrix}}{\langle 1 \quad -1 \rangle \begin{Bmatrix} 1 \\ -1 \end{Bmatrix}} = 1 \qquad [1.92]$$

The search direction can then be written as follows:

$$\{S\}_{k+1} = \nabla\{f(x)\}_{k+1} + \beta_{k+1}\{S\}_k \qquad [1.93]$$

The search direction for this iteration is:

$$S_1 = \begin{Bmatrix} -1 \\ -1 \end{Bmatrix} + 1\begin{Bmatrix} 1 \\ -1 \end{Bmatrix} = \begin{Bmatrix} 0 \\ -2 \end{Bmatrix} \qquad [1.94]$$

The iterative equation becomes:

$$x_{(2)} = x_{(1)} - \alpha_2 S_1 \qquad [1.95]$$

When considering the forward step as a variable, we obtain:

$$x_{(2)} = \begin{Bmatrix} -1 \\ 1 \end{Bmatrix} - \alpha_2 \begin{Bmatrix} 0 \\ -2 \end{Bmatrix} = \begin{Bmatrix} -1 \\ 1 + 2\alpha_2 \end{Bmatrix} \qquad [1.96]$$

We calculate the forward step by substituting the optimization variables in the objective function formulation as follows:

$$f(\alpha_2) = -1 - (1 + 2\alpha_2) + 2(-1)^2 + 2(-1)(1 + 2\alpha_2) + (1 + 2\alpha_2)^2 \qquad [1.97]$$

The equation obtained can be simplified into the following form:

$$f(\alpha_2) = -1 - 2\alpha_2 + 4\alpha_2^2 \qquad [1.98]$$

The derivative of the iterative equation with respect to the current step gives:

$$\frac{\partial f(\alpha_2)}{\partial \alpha_2} = -2 + 8\alpha_2 = 0 \Rightarrow \alpha_2 = \frac{1}{4} \qquad [1.99]$$

The new value is:

$$x_{(2)} = \begin{Bmatrix} -1 \\ 1.5 \end{Bmatrix} \qquad [1.100]$$

At the optimal point, we find that the derivative is null:

$$\nabla f(\mathbf{x}_{(2)}) = \begin{Bmatrix} 0 \\ 0 \end{Bmatrix} \qquad [1.101]$$

Figure 1.26 shows the coordinates of the new value at iteration (2).

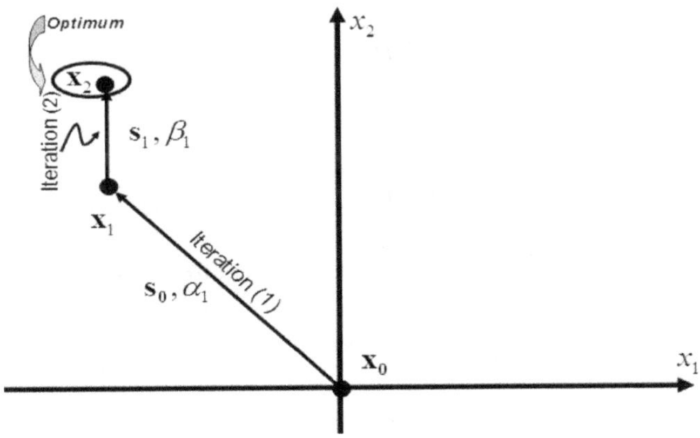

Figure 1.26. *Iterations 1 and 2 using CGM*

Newton's method

The second-order methods are generally efficient to find the optimal solutions. They use calculations of second-order gradients. We calculate the first-order derivative:

$$S_0 = \nabla f(\mathbf{x_0}) = \begin{Bmatrix} 1 \\ -1 \end{Bmatrix} \qquad [1.102]$$

Then, we calculate the Hessian matrix as follows:

$$H(x_O) = \begin{bmatrix} \dfrac{\partial^2 f(x_0)}{\partial x_1^2} & \dfrac{\partial^2 f(\mathbf{x_0})}{\partial x_1 \partial x_2} \\ \dfrac{\partial^2 f(x_0)}{\partial x_2 \partial x_1} & \dfrac{\partial^2 f(x_0)}{\partial x_2^2} \end{bmatrix} = \begin{bmatrix} 4 & 2 \\ 2 & 2 \end{bmatrix} \qquad [1.103]$$

Next, the inverse of this matrix (see Appendix 1) is then:

$$H(x_O)^{-1} = \begin{bmatrix} \dfrac{1}{2} & -\dfrac{1}{2} \\ -\dfrac{1}{2} & 1 \end{bmatrix} \qquad [1.104]$$

Hence, the search direction can be expressed by:

$$d_{k+1} = x_{k+1} - x_k = -[H(x_k)]^{-1} \nabla f(x_k) \qquad [1.105]$$

and the iterative equation is expressed as:

$$x_1 = x_0 - [H(x_0)]^{-1} \nabla f(x_0) \qquad [1.106]$$

Considering the iterative equation, we obtain:

$$x_1 = \begin{Bmatrix} 0 \\ 0 \end{Bmatrix} - \begin{bmatrix} \dfrac{1}{2} & -\dfrac{1}{2} \\ -\dfrac{1}{2} & 1 \end{bmatrix} \begin{Bmatrix} 1 \\ -1 \end{Bmatrix}^* \qquad [1.107]$$

The new value is thus:

$$x_1 = \begin{Bmatrix} -1 \\ \frac{3}{2} \end{Bmatrix} \qquad [1.108]$$

At the optimal point, we find that the derivative is null:

$$\nabla f(\mathbf{x}_1) = \begin{Bmatrix} 0 \\ 0 \end{Bmatrix} \qquad [1.109]$$

NM converges after a single iteration.

Modified Newton's method

This method only uses the Hessian matrix for the first iteration. We obtain the inverse of the Hessian matrix as before (see equation [1.104]). However, we introduce the step to the iterative equation as follows:

$$x_{(1)} = x_{(0)} - \alpha\left[H(x_0)\right]^{-1} \nabla f(x_0) \qquad [1.110]$$

When considering the forward step as a variable, we obtain:

$$x_{(1)} = \begin{Bmatrix} 0 \\ 0 \end{Bmatrix} - \alpha \begin{bmatrix} 0.5 & -0.5 \\ -0.5 & 1 \end{bmatrix} \begin{Bmatrix} 1 \\ -1 \end{Bmatrix} = \begin{Bmatrix} -\alpha \\ 1.5\alpha \end{Bmatrix} \qquad [1.111]$$

We calculate the forward step by substituting the optimization variables in the objective function formulation as follows:

$$f(\alpha) = -\alpha - 1.5\alpha + 2(-\alpha)^2 + 2(-\alpha)(1.5\alpha) + (1.5)^2\alpha^2 \qquad [1.112]$$

The obtained equation can be simplified into the following form:

$$f(\alpha) = -2.5\alpha + 1.25\alpha^2 \qquad [1.113]$$

The derivative of the iterative equation with respect to the current step gives:

$$\frac{\partial f(\alpha)}{\partial \alpha} = 2.5\alpha - 2.5 = 0 \Rightarrow \alpha = 1 \qquad [1.114]$$

The new value is:

$$x_{(1)} = \begin{Bmatrix} -1 \\ 1.5 \end{Bmatrix} \quad [1.115]$$

At the optimal point, we find that the derivative is null:

$$\nabla f(\mathbf{x}_1) = \begin{Bmatrix} 0 \\ 0 \end{Bmatrix} \quad [1.116]$$

Figure 1.27 shows the coordinates of the new value at iteration (1).

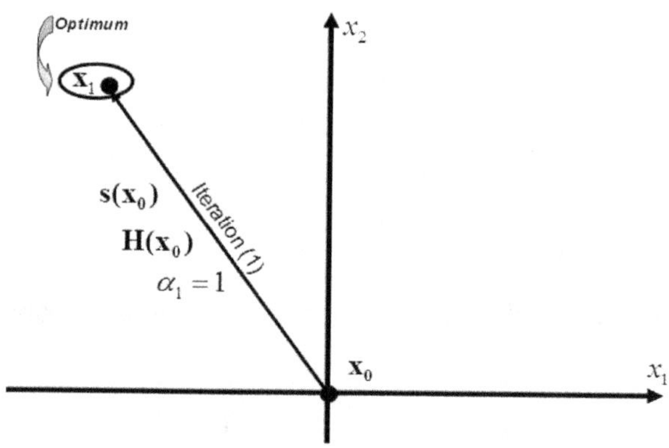

Figure 1.27. *Iteration (1) using NM and MNM*

The initial value of the objective is $f(\mathbf{x}_0) = 0$, while the minimal value is $f(\mathbf{x}^*) = -0.5$.

This equation can be solved using iterative techniques.

Method	*DGM*	*CGM*	*NM*	*MNM*
Iterations	6	2	1	1

Table 1.3. *Number of iterations for the different methods*

Table 1.3 shows a comparison between the four methods used. For this example, the CGM is very efficient relative to the DGM. Although the second-order methods only require one iteration, the evaluation of the Hessian matrix is a potential issue during optimization.

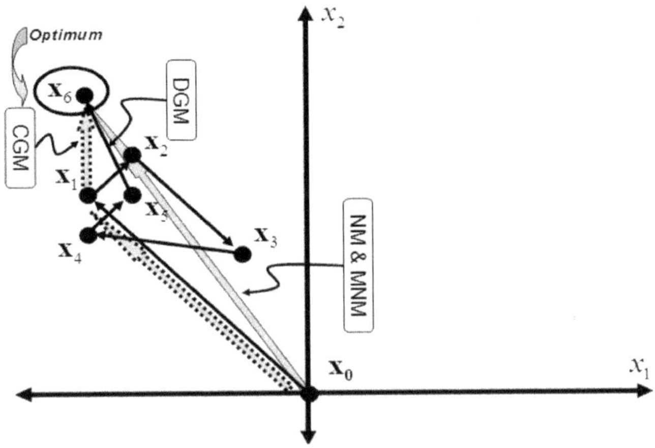

Figure 1.28. *Different iterations for the different methods*

Figure 1.28 shows the different iterations (trajectories) for the different methods used.

1.4.5. *Constrained methods*

In general, constrained optimization methods can be divided into two categories: direct methods and transformation methods.

1.4.5.1. *Direct methods*

The direct methods use linear or nonlinear programming techniques.

1.4.5.1.1. SLP method

The sequential linear programming or successive linear programming (SLP) is an optimization technique for approximately solving nonlinear optimization problems method. When starting at some estimate of the optimal solution, the method is based on solving a sequence of first-order approximations (i.e. linearizations) of the model. The linearizations are linear programming problems, which can be solved efficiently.

1.4.5.1.2. SQP method

SQP methods solve a sequence of optimization subproblems, each of which optimizes a quadratic model of the objective function subject to a linearization of the constraints. If the problem is unconstrained, then the method reduces to NM for finding a point where the gradient of the objective vanishes. If the problem has only equality constraints, then the method is equivalent to applying NM to the first-order optimality conditions of the problem. SQP methods have been implemented in many packages such as MATLAB.

1.4.5.2. *Transformation methods*

These methods transform constrained problems into new formulations by increasing the number of variables.

1.4.5.2.1. Lagrange method

Joseph-Louis Compte de Lagrange (1736–1813) was a French mathematician, born in Italy. He worked on isoparametric problems, founded calculus of variations and contributed to calculus of probabilities, the theory of equations and group theory.

The Lagrange method uses Lagrange multipliers for equality constraints, and adjustment variables or slack variables for inequality constraints. The Lagrange formulation of a constrained optimization problem can be written using the following form:

$$L(\mathbf{x},\boldsymbol{\lambda},\boldsymbol{\mu}) = f(\mathbf{x}) + \sum_{j=1}^{L} \lambda_j h_j(\mathbf{x}) + \sum_{k=1}^{m} \mu_k g_k(\mathbf{x}) \qquad [1.117]$$

The Lagrange formulation is a function of $\mathbf{x}$, $\boldsymbol{\lambda}$ and $\boldsymbol{\mu}$. To solve this problem, we use unconstrained methods.

ILLUSTRATIVE EXAMPLE 1.5.–

Consider an optimization problem of the following form:

$$\min f(\mathbf{x}) = -3x_1 - x_2,$$
$$\text{subject to} \begin{cases} -6x_1 + 5x_2 \le 30 \\ -7x_1 + 12x_2 \le 84 \\ x_2 \le 9 \\ 19x_1 + 14x_2 \le 266 \\ x_1 \le 10 \\ 4x_1 - 7x_2 \le 28 \\ x_1 \ge 0, x_2 \ge 0 \end{cases} \qquad [1.118]$$

– classify this optimization problem;

– write the Lagrange formulation of this problem;

– model geometrically all functions (objective and constraints);

– find the optimal solution.

SOLUTION.–

The classification of this optimization problem is a constrained, linear, single-objective problem.

The Lagrange formulation is then written for the constrained optimization problem, where its general form is expressed as follows:

$$\begin{array}{ll} \min & :f(\mathbf{x}) \\ \text{subject to} & :h_i(\mathbf{x}) = 0 \\ \text{and} & :g_j(\mathbf{x}) \le 0 \end{array} \qquad [1.119]$$

Hence, the problem being studied here can be reformulated into the following form:

$$\min f(\mathbf{x}) = -3x_1 - x_2,$$
$$\text{subject to} \begin{cases} -6x_1 + 5x_2 - 30 \le 0 \\ -7x_1 + 12x_2 - 84 \le 0 \\ x_2 - 9 \le 0 \\ 19x_1 + 14x_2 - 266 \le 0 \\ x_1 - 10 \le 0 \\ 4x_1 - 7x_2 - 28 \le 0 \\ -x_1 \le 0, -x_2 \le 0 \end{cases} \qquad [1.120]$$

The constraints are inequality constraints. They are expressed as follows:

$$\begin{aligned} g_1(x) &= -6x_1 + 5x_2 - 30 \le 0 \\ g_2(x) &= -7x_1 + 12x_2 - 84 \le 0 \\ g_3(x) &= x_2 - 9 \le 0 \\ g_4(x) &= 19x_1 + 14x_2 - 266 \le 0 \\ g_5(x) &= x_1 - 10 \le 0 \\ g_6(x) &= 4x_1 - 7x_2 - 28 \le 0 \\ g_7(x) &= -x_1 \le 0 \\ g_8(x) &= -x_2 \le 0 \end{aligned} \qquad [1.121]$$

To model this problem geometrically, we model all functions as straight lines. The intersection of the first straight line with the Cartesian axes gives:

$$g_1(x) = -6x_1 + 5x_2 - 30 = 0 \;\Rightarrow\; \begin{Bmatrix} P_1(0,6) \\ P_2(-5,0) \end{Bmatrix} \qquad [1.122]$$

The intersection of the second straight line with the Cartesian axes gives:

$$g_2(x) = -7x_1 + 12x_2 - 84 = 0 \;\Rightarrow\; \begin{Bmatrix} P_1(0,7) \\ P_2(-12,0) \end{Bmatrix} \qquad [1.123]$$

The third constraint $g_3(x) = x_2 - 9 = 0$ is parallel to the x_1 axis and the intersection is at the point $P(0,9)$. The intersection of the fourth straight line with the Cartesian axes gives:

$$g_4(x) = 19x_1 + 14x_2 - 266 = 0 \quad \Rightarrow \quad \begin{Bmatrix} P_1(0,19) \\ P_2(14,0) \end{Bmatrix} \qquad [1.124]$$

The fifth constraint $g_5(x) = x_1 - 10 = 0$ is parallel to the x_2 axis and the intersection is at the point $P(10,0)$. The intersection of the sixth straight line with the Cartesian axes gives:

$$g_6(x) = 4x_1 - 7x_2 - 28 = 0 \quad \Rightarrow \quad \begin{Bmatrix} P_1(0,-4) \\ P_2(7,0) \end{Bmatrix} \qquad [1.125]$$

The seventh constraint $g_7(x) = -x_1 = 0$ coincides with the x_2 axis and the eighth constraint $g_8(x) = -x_2 = 0$ coincides with the x_1 axis.

To model the iso-values of the objective $f(\mathbf{x}) = -3x_1 - x_2$, we consider several straight lines (several levels). The intersection of the first straight line with the Cartesian axes gives:

$$f(x) = -3x_1 - x_2 = 0 \quad \Rightarrow \quad \begin{Bmatrix} P_1(0,0) \\ P_2(1,-3) \end{Bmatrix} \qquad [1.126]$$

The intersection of the second straight line with the Cartesian axes gives:

$$f(x) = -3x_1 - x_2 = 1 \quad \Rightarrow \quad \begin{Bmatrix} P_1(0,-1) \\ P_2(-\frac{1}{3},0) \end{Bmatrix} \qquad [1.127]$$

Figure 1.29 shows the point of the optimal solution $\mathbf{x}^* = (10,5.4)$ where the maximal value of the objective is $f(\mathbf{x}^*) = -35.4$.

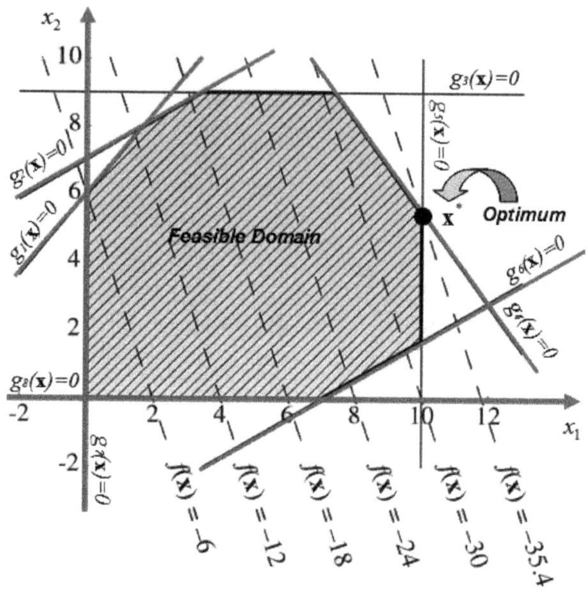

Figure 1.29. *Geometrical modeling of the optimal point*

1.4.5.2.2. Optimality criteria method

The optimality criteria (OC) method uses the first-order derivatives of the Lagrange formulation. The necessary first-order derivative conditions giving the critical points are written as follows:

$$\frac{\partial L(\mathbf{x},\boldsymbol{\lambda},\boldsymbol{\mu})}{\partial \mathbf{x}} = \frac{\partial f(\mathbf{x})}{\partial \mathbf{x}} + \sum_{j=1}^{j} \lambda_j \frac{\partial h_j(\mathbf{x})}{\partial \mathbf{x}} + \sum_{k=1}^{k} \mu_k \frac{\partial g_k(\mathbf{x})}{\partial \mathbf{x}} = 0 \qquad [1.128]$$

$$\frac{\partial L(\mathbf{x},\boldsymbol{\lambda},\boldsymbol{\mu})}{\partial \boldsymbol{\lambda}} = h_j(\mathbf{x}) = 0 \qquad [1.129]$$

$$\frac{\partial L(\mathbf{x},\boldsymbol{\lambda},\boldsymbol{\mu})}{\partial \boldsymbol{\mu}} = g_k(\mathbf{x}) = 0 \qquad [1.130]$$

The solution of this problem leads to an optimum which is local and can be global.

ILLUSTRATIVE EXAMPLE 1.6.–

Consider a constrained optimization problem of the following form:

$$\begin{aligned} &\min \quad :f(\mathbf{x}) = x_1^2 + x_1 x_2, \\ &\text{subject to:} g(\mathbf{x}) = -x_1^2 x_2 + 300/\pi \leq 0, \\ &\qquad\qquad : x_i \geq 0, i = 1, 2. \end{aligned} \qquad [1.131]$$

– classify the optimization problem;

– find geometrically the optimal solution;

– find the minimal value using the optimality criteria method.

SOLUTION.–

The classification of this optimization problem is a constrained, nonlinear and single-objective problem.

To model this problem geometrically, we start with the iso-values of the objective function. For the straight line $f(\mathbf{x}) = x_1^2 + x_1 x = 10$, we obtain the following points: $P_1(1,9)$, $P_2(5,-3)$ and $P_3(10,-9)$. Next, for the straight line $f(\mathbf{x}) = x_1^2 + x_1 x = 20$, we obtain the following points: $P_1(1,19)$, $P_2(5,-1)$ and $P_3(10,-8)$.

To model the first constraint $g_1(\mathbf{x}) = -x_1^2 x_2 + 300/\pi = 0$, we obtain the following points: $P_1(10,0.95)$, $P_2(10,3.09)$ and $P_3(15,0.42)$ which corresponds to objective function values of: $f_1(\mathbf{x}) = 109.54$, $f_2(\mathbf{x}) = 40.44$ and $f_3(\mathbf{x}) = 232.35$. The second constraint $g_2(x) = -x_1 = 0$ coincides with the x_2 axis and the third constraint $g_3(x) = -x_2 = 0$ coincides with the x_1 axis. Figure 1.30 shows a representation of this optimization problem.

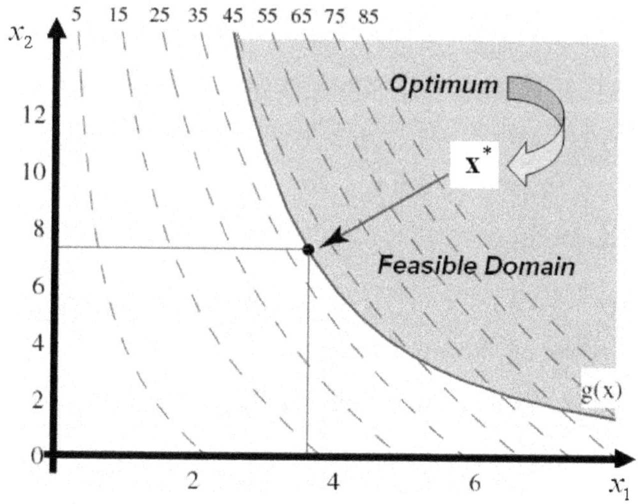

Figure 1.30. *Geometrical modeling of the studied optimization problem*

We find that the problem is nonlinear, meaning that an iterative method is required to find the optimal solution.

To use the optimality criteria method, we rewrite the constraints considering the standard forms:

$$\begin{aligned} g_1(x) &= -x_1^2.x_2 + \frac{300}{\pi} \leq 0 \\ g_2(x) &= -x_1 \leq 0 \\ g_3(x) &= -x_2 \leq 0 \end{aligned} \qquad [1.132]$$

The Lagrange formulation can be written using the following form:

$$L = x_1^2 + x_1.x_2 + \mu_1(-x_1^2 x_2 + \frac{300}{\pi}) + \mu_2(-x_1) + \mu_3(-.x_2) \qquad [1.133]$$

By differentiating this formulation, we obtain:

$$\frac{\partial L}{\partial x_1} = 2x_1 + x_2 - 2\mu_1 x_1 x_2 - \mu_2 = 0 \qquad [1.134]$$

$$\frac{\partial L}{\partial x_2} = x_1 - \mu_1 x_1^2 - \mu_3 = 0 \qquad [1.135]$$

$$\frac{\partial L}{\partial \mu_1} = -x_1^2 x_2 + \frac{300}{\pi} = 0 \qquad [1.136]$$

$$\frac{\partial L}{\partial \mu_2} = -x_1 = 0 \qquad [1.137]$$

$$\frac{\partial L}{\partial \mu_3} = -x_2 = 0 \qquad [1.138]$$

Assuming that the two variables are not null ($x_1 \neq 0$ and $x_2 \neq 0$), we obtain ($\mu_2 = \mu_3 = 0$). Equation [1.134] becomes:

$$2x_1 + x_2 - 2\mu_1 x_1 x_2 = 0 \qquad [1.139]$$

which leads to:

$$\mu_1 = \frac{2x_1 + .x_2}{2x_1 x_2} \qquad [1.140]$$

When substituting this result into equation [1.135], we obtain:

$$x_1 - \mu_1 .x_1^2 = 0 \Rightarrow x_1 - \left(\frac{2x_1 + .x_2}{2x_1 x_2} \right) x_1^2 = 0 \Rightarrow x_2 = 2x_1 \qquad [1.141]$$

which leads to:

$$x_2 = 2x_1 \qquad [1.142]$$

When substituting this result into equation [1.136], we obtain:

$$-x_1^2 x_2 + \frac{300}{\pi} = 0 \Rightarrow -x_1^2 (2x_1) + \frac{300}{\pi} = 0 \qquad [1.143]$$

which leads to the optimal solution $x_1 = 3.627$ and $x_2 = 7.2556$. The value of the objective function is then $f(\mathbf{x}^*) = 39.47$. Figure 1.30 shows the geometrical interpretation of this optimization problem.

ILLUSTRATIVE EXAMPLE 1.7.–

Consider an optimization problem of the following form:

$$\begin{aligned} &\text{min} \quad :Cost(b,h) = 12b + 10h - 2b^2 - bh - h^2 \\ &\text{subject to:} b + h \geq 4 \end{aligned} \quad [1.144]$$

– classify the optimization problem;

– find geometrically the optimal solution;

– find the minimal value using the optimality criteria method.

SOLUTION.–

The classification of this optimization problem is a constrained, nonlinear and single-objective problem.

To geometrically model this problem, we rewrite the inequality constraint in the standard form:

$$g_1(x) = 4 - b - h \leq 0 \quad [1.145]$$

The boundary of this constraint corresponds to a straight line. The intersection of this straight line with the Cartesian axes gives:

$$g_1(x) = 4 - b - h = 0 \quad \Rightarrow \quad \begin{Bmatrix} P_1(0,4) \\ P_2(4,0) \end{Bmatrix} \quad [1.146]$$

Next, we model the iso-values of the objective function (several levels). At the level $Cost(b,h) = 12b + 10h - 2b^2 - bh - h^2 = 0$, the intersection with the horizontal axis gives:

$$b = 0 \Rightarrow 10h - h^2 = 0 \Rightarrow h = 0, h = 10 \quad [1.147]$$

The intersection with the vertical axis gives:

$$h = 0 \Rightarrow 12b - 2b^2 = 0 \Rightarrow b = 0, b = 6 \qquad [1.148]$$

At the level $Cost(b,h) = 12b + 10h - 2b^2 - bh - h^2 = 10$, the intersection with the horizontal axis gives:

$$b = 0 \Rightarrow 10h - h^2 = 10 \Rightarrow h = 1.127, h = 8.87 \qquad [1.149]$$

The intersection with the vertical axis gives:

$$h = 0 \Rightarrow 12b - 2b^2 = 0 \Rightarrow b = 1, b = 5 \qquad [1.150]$$

At the level $Cost(b,h) = 12b + 10h - 2b^2 - bh - h^2 = 20$, the intersection with the horizontal axis gives:

$$b = 0 \Rightarrow 10h - h^2 = 20 \Rightarrow h = 2.76, h = 7.23 \qquad [1.151]$$

The intersection with the vertical axis gives:

$$b = 1 \Rightarrow 12 + 10h - 2 - h - h^2 = 20 \Rightarrow h = 1.29, h = 7.7 \qquad [1.152]$$

which leads to the optimal solution $b =$ 1.5 and $h = 2.5$. The value of the objective function is then $f(b,h)$ = 28.5. Figure 1.31 shows the geometrical interpretation of this optimization problem.

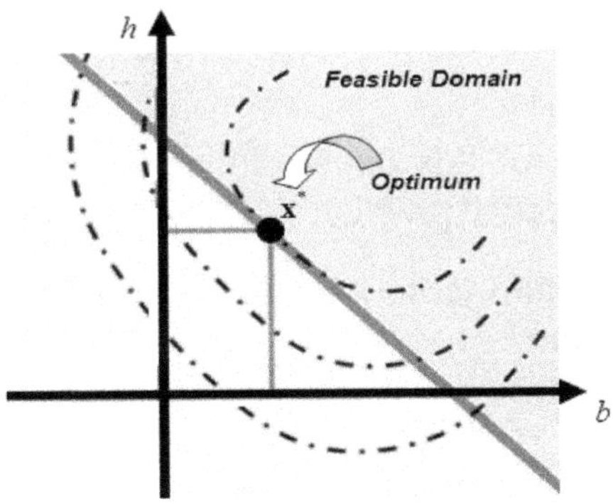

Figure 1.31. *Geometric model of the optimization problem*

The Lagrange formulation can be written using the following form:

$$L(b,h,\mu)=12b+10h-2b^2-bh-h^2+\mu[4-b-h] \qquad [1.153]$$

When differentiating this formulation, we obtain:

$$\frac{\partial L}{\partial b}=0 \Rightarrow 12-4b-h-\mu=0 \qquad [1.154]$$

$$\frac{\partial L}{\partial h}=0 \Rightarrow 10-b-2h-\mu=0 \qquad [1.155]$$

$$\frac{\partial L}{\partial \mu}=0 \Rightarrow 4-b-h=0 \qquad [1.156]$$

We have three linear equations with three unknowns. The analytic solution gives b = 1.5, h = 2.5 and μ = 3.5. The minimal value of the objective function is thus min $Cost$ = 28.5.

ILLUSTRATIVE EXAMPLE 1.8.–

Consider an optimization problem in the following form:

$$\begin{array}{ll} \text{min} & : f(\mathbf{x}) = 2x_1^2 - 2x_1 + \frac{1}{2}x_2^2 - 3x_2 \\ \text{subject to} & : -(x_1-1)^2 - x_2^2 + 2x_2 \geq 1 \\ \text{and} & : x_1 \geq 0, x_2 \geq 0 \end{array} \qquad [1.157]$$

– classify the optimization problem;

– find the minimal value using the optimality criteria method.

SOLUTION.–

The classification of this optimization problem is a constrained, nonlinear, single-objective problem.

To use the optimality criteria method, we rewrite the constraints using the standard form:

$$\begin{array}{l} g_1(\mathbf{x}) = (x_1-1)^2 + x_2^2 - 2x_1 + 1 \leq 0 \\ g_2(x) = -x_1 \leq 0 \\ g_3(x) = -x_2 \leq 0 \end{array} \qquad [1.158]$$

The Lagrange formulation can be written as follows:

$$L = 2x_1^2 - 2x_1 + \frac{1}{2}x_2^2 - 3x_2 + \mu_1\left[(x_1-1)^2 + x_2^2 - 2x_1 + 1\right] + \mu_2(-x_1) + \mu_3(-x_2) \quad [1.159]$$

When differentiating this formulation, we obtain:

$$\frac{dL}{\partial x_1} = 4x_1 - 2 + 2\mu_1(x_1 - 1) - 2\mu_1 - \mu_2 = 0 \quad [1.160]$$

$$\frac{dL}{\partial x_2} = x_2 - 3 + 2\mu_1 x_2 - \mu_3 = 0 \quad [1.161]$$

$$\frac{dL}{\partial \mu_1} = (x_1 - 1)^2 + x_2^2 - 2x_1 + 1 = 0 \quad [1.162]$$

$$\frac{\partial L}{\partial \mu_2} = -x_1 = 0 \quad [1.163]$$

$$\frac{\partial L}{\partial \mu_3} = -x_2 = 0 \quad [1.164]$$

Assuming that the two variables are not null ($x_1 \neq 0$ and $x_2 \neq 0$), we obtain ($\mu_2 = \mu_3 = 0$). Equation [1.160] becomes:

$$4x_1 - 2 + 2\mu_1(x_1 - 2) = 0 \quad [1.165]$$

which leads to:

$$\mu_1 = \frac{2 - 4x_1}{2(x_1 - 2)} \quad [1.166]$$

When substituting this result into equation [1.161], we obtain:

$$x_2 - 3 + 2\mu_1 x_2 = 0 \quad [1.167]$$

which leads to:

$$x_2 = \sqrt{2x_1 - 1 - (x_1 - 1)^2} \qquad [1.168]$$

When substituting this result into equation [1.162], we obtain:

$$\sqrt{2x_1 - 1 - (x_1 - 1)^2} - 3 + 2\left(\frac{2 - 4x_1}{2(x_1 - 2)}\right)\sqrt{2x_1 - 1 - (x_1 - 1)^2} = 0 \qquad [1.169]$$

which leads to the optimal solution $x_1 = 1$, $x_2 = 1$ and $\mu_1 = 1$. The geometric solution is only effective for bivariable problems with linear functions.

1.5. Sensitivity analysis

1.5.1. *Importance of sensitivity*

Sensitivity analysis of a mathematical model consists of studying the variation impact of the input parameters on the output parameters of the model. The sensitivity of a function (or indeed the derivative of this function) with respect to several variables gives the influence of each variable on the studied function. The sensitivity calculation can be explicit or implicit [KHA 11].

ILLUSTRATIVE EXAMPLE 1.9.–

Figure 1.32 shows a cantilever beam of a rectangular cross-section ($b \times h$) subjected to a static force F and fixed at the other end.

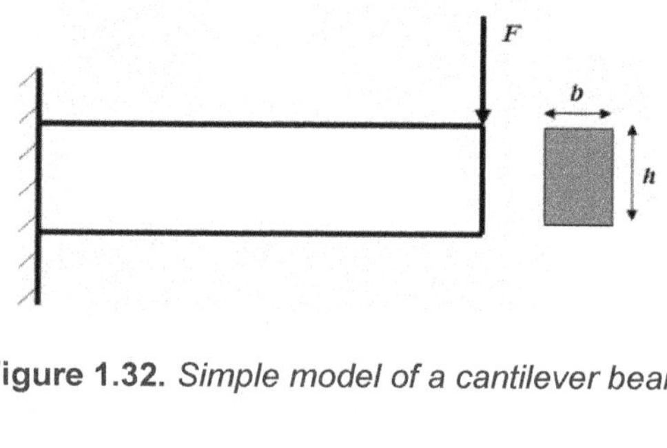

Figure 1.32. *Simple model of a cantilever beam*

The maximal mechanical stress can be found at the end of the beam and can be mathematically calculated as follows:

$$\sigma_{\max} = \frac{M_{\max}}{I} \times \frac{h}{2} \quad [1.170]$$

where I is the inertia moment for a rectangular cross-section and can be calculated using the following relationship:

$$I = \frac{b.h^3}{12} \quad [1.171]$$

The derivatives of the maximal mechanical stress with respect to the two dimensions show that the beam height has much more of an effect than the beam width ($\partial I/\partial h >> \partial I/\partial b$). In this example, we have an explicit model because the derivative can be calculated analytically. In contrast, Figure 1.33 shows the same beam, but with transverse holes drilled through it, which leads to an infinite number of cross-sections.

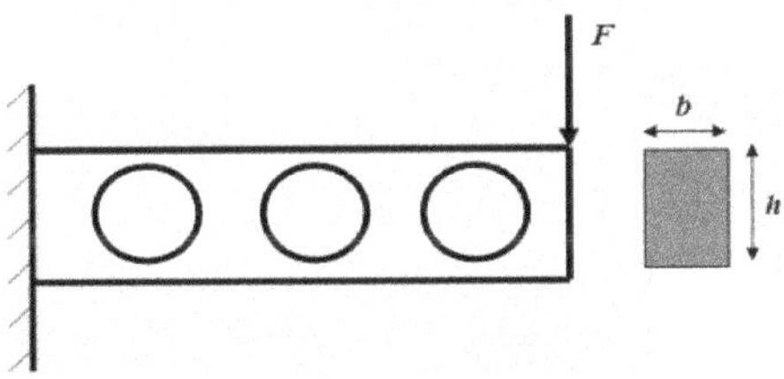

Figure 1.33. *Implicit model of a cantilever beam*

In order to calculate the derivative of this implicit model, we have to use numerical methods to carry out the sensitivity analysis.

1.5.2. *Sensitivity methods*

Sensitivity analysis is an important tool during the development, construction or use of a mathematical model. In fact, when studying how a model responds to variations in its input variables, sensitivity analysis can answer a number of questions. Sensitivity analysis can be carried out in three ways: numerically, analytically and semi-analytically [KHA 03]. We are interested in the numerical method of finite differences, which is very simple to implement.

Using the finite difference (FD) method, we calculate the first derivative $df(x_i)/dx_i$ with the following techniques:

1) Forward finite difference (FFD):

$$\frac{df(x_i)}{dx_i} \cong \left[\frac{\Delta f(x_i)}{\Delta x_i}\right]_{FFD} = \frac{f(x_i + \Delta x_i) - f(x_i)}{\Delta x_i} \quad [1.172]$$

2) Backward finite difference (BFD):

$$\frac{df(x_i)}{dx_i} \cong \left[\frac{\Delta f(x_i)}{\Delta x_i}\right]_{BFD} = \frac{f(x_i) - f(x_i - \Delta x_i)}{\Delta x_i} \quad [1.173]$$

3) Central finite difference (CFD):

$$\frac{df(x_i)}{dx_i} \cong \left[\frac{\Delta f(x_i)}{\Delta x_i}\right]_{CFD} = \frac{f(x_i + \Delta x_i) - f(x_i - \Delta x_i)}{2\Delta x_i} \quad [1.174]$$

The choice of the increment Δx_i to achieve a good accuracy is difficult. Figure 1.34 shows that the value of the derivative varies as a function of Δx_i.

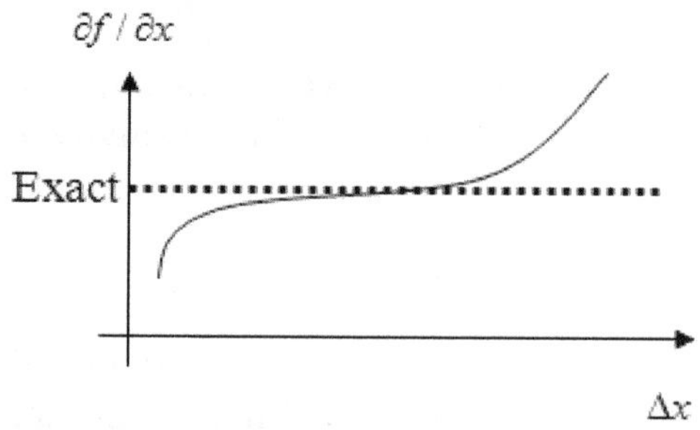

Figure 1.34. *Numerical derivative as a function of the increment*

ILLUSTRATIVE EXAMPLE 1.10.–

Considering a function $f(x) = 4x^2 - 8x + 13$, we test the finite difference precision by calculating the sensitivity or the derivatives of this function at the point $x = 2.5$.

Using the FFD technique, the change of Δx_i can lead to the exact value of the derivative $\partial f(x)/\partial x$ (Figure 1.35).

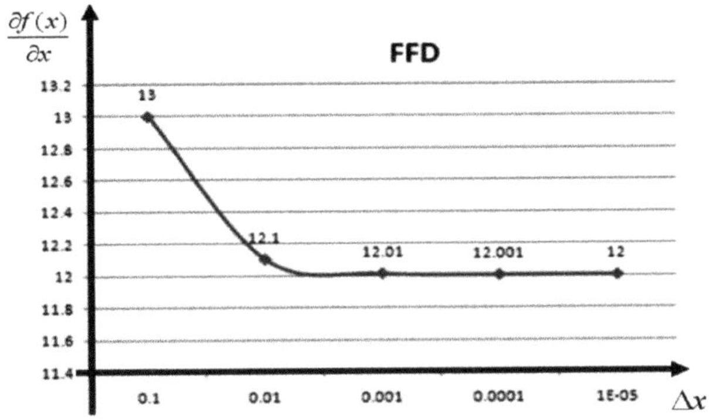

Figure 1.35. *Precision curve using FFD technique*

Using the BFD technique, the change of Δx_i can lead to the exact value of the derivative $\partial f(x)/\partial x$ (Figure 1.36).

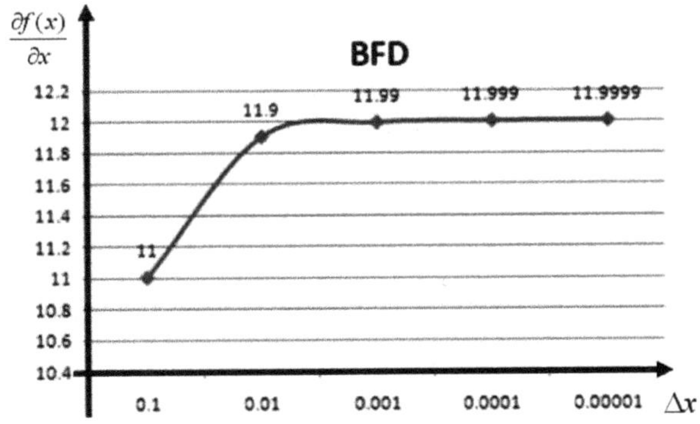

Figure 1.36. *Precision curve using BFD technique*

Using the CFD technique, the change of Δx_i can lead to the exact value of the derivative $\partial f(x)/\partial x$ (Figure 1.37).

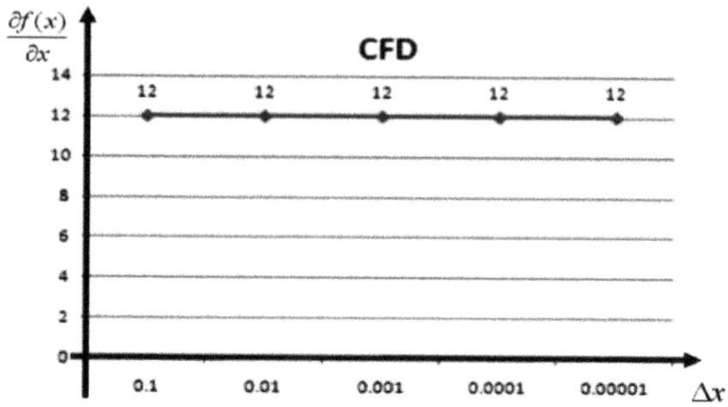

Figure 1.37. *Precision curve using CFD technique*

We find that the CFD technique gives exact values for all values of the increment Δx_i. Figure 1.38 shows the algorithm for finite difference calculations of derivatives with high precision, ε.

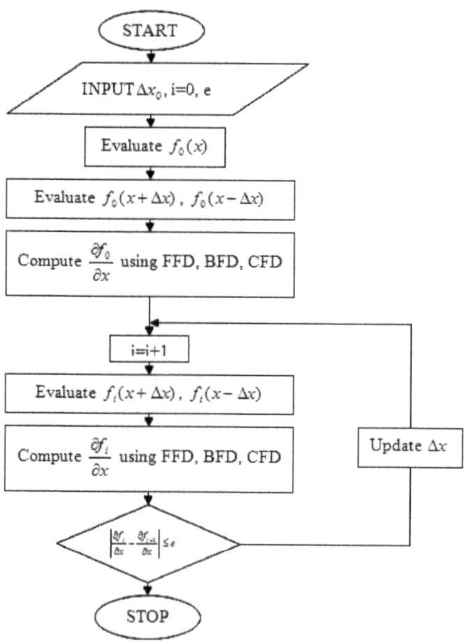

Figure 1.38. *Algorithm of precision, ε, with finite difference methods*

Furthermore, considering the precision advantage of the CFD technique, we can also calculate the second derivative as a function of a single variable using the following formula:

$$\frac{d^2 f}{dx_i^2} = \frac{f(x_i + 2\Delta x_i) - 2f(x_i + \Delta x_i) + f(x_i)}{\Delta x_i^2} \qquad [1.175]$$

In the same way, we can calculate the second derivative as a function of two variables as follows:

$$\frac{d^2 f}{dx_i dx_j} = \dots = \frac{f(x_i + \Delta x_i, x_j + \Delta x_j) - f(x_i + \Delta x_i, x_j) - f(x_i, x_j + \Delta x_j) + f(x_i, x_j)}{\Delta x_i \Delta x_j} \qquad [1.176]$$

The equations for the first-order derivatives can be used in first-order optimization methods (DGM and CGM). In contrast, the equations for the second-order derivatives can be used in the second-order optimization methods (NM and MNM) [KHA 11].

1.6. Conclusion

In this chapter, several mathematical and numerical tools have been presented as basic concepts to be used in the following chapters. Geometric models in biomechanics are very complicated and require a numerical simulation (for example, FEM) to solve the studied problem. Understanding the optimization technology is an essential stage to apply reliability analysis in biomechanics. Additionally, sensitivity analysis can be carried out during the optimization process and reliability analysis in order to identify the influence of input parameters on the output responses.

2

Reliability Concept

2.1. Introduction

2.1.1. *Preamble*

The noun “reliability” comes from the adjective reliable, which in itself is derived from the verb to rely (or to trust) and the word “ability”. The technical concept of “ability of a device to perform a required function under given conditions, for a given period” [LIT 97]. Having become an important sales argument, reliability remains a major concern for engineers to study.

2.1.2. *Reliability history*

The safety of persons is a crucial issue in all societies, particularly with the progress made in the field of science. To this end, the failure of a structure could have catastrophic consequences on the human being and the environment.

Since ancient times, man has tried to develop regulations to ensure a certain level of safety of structures. The Hammurabi code is considered to be the first known legal code governing the acceptance of risk. It was made on the initiative of the King of Babylon around 1730 BC. This code is designed for the regulation of everyday life of citizens to organize the relationships between the individuals. This code contains several articles, which include articles 229 and 230, which aimed to ensure a certain level of security in houses.

Articles 229 and 230 are presented as follows [FIN 83]:

– Article 229: "if a builder has built a house for someone, but he has not reinforced his construction and the house that he has built collapses and kills the owner of the house, the builder will be killed."

– Article 230: "if it is a child of the house owner who is killed, a child of the builder shall be killed."

To ensure an acceptable level of safety for users, the code of Hammurabi requires the transfer of a risk proportional to the designer. Even if this code seems inhuman nowadays, nevertheless it remains a direct application of the relationship "'probability, gravity". In fact, Farmer [FAR 63] defines risk by the existence of a dreaded event that can cause the structure failure, and by the gravity of it on the structure using the following relation:

$$Risk = Probability \times hazard \qquad [2.1]$$

The Code of Hammurabi aims to decrease the risk of the event feared by the customer by increasing the consequences for whoever ensures the responsibility of the construction.

Man's knowledge has improved by trial-and-error since the times of Hammurabi. The work of Leonardo da Vinci (1452–1519) in the construction of cathedrals, and the work of Galileo (1564–1642) on optimizing the resistance of a beam to fracture, laid the foundations of new perspectives in the research of laws governing uncertainty in structures.

Industrial interest in reliability is not new, for a number of reasons. Objectives for extending the lifetimes of vacuum tubes date back to

1906, and such work continued in the steam-locomotive sector during World War I, with the evaluation of the lifetimes of rolling bearings. At that time, only the concept of lifetimes was associated with reliability. During World War II, with the integration of electronics in equipment, reliability became an urgent requirement: being prone to many failures, these systems represented a large investment in terms of maintenance costs. It was during the 1950s that reliability emerged as a field of its own. It was during this time that the Americans were the first to realize that reliability put a limitation on the complexity of systems. In 1962, the first method for estimating reliability appeared in the form of a military handbook. In France, 1962 was also the year of the first tangible manifestation of the realization of this problem, with the creation of the "Centre of Reliability" at CNET, at the same time as the term reliability was accepted by the French Academy of Sciences. However, it was not until 1970 that a method of estimation saw the light of day in France (*Recueil de Fiabilité RdF 70*). Starting from the 1970s onward, reliability analysis of systems became recognized, and then an obligatory activity in several sectors such as nuclear power, aerospace, automotive, aeronautics and defense.

2.1.3. *Reliability definition*

The definition adopted by [MAR 74] for the neologism reliability was "reliability is a characteristic of a device expressed by the probability that a required function is performed under given conditions, during a given time".

Reliability can be characterized by the four following elements:

– probability: it is defined by the ratio between the number of favorable cases to the realization of an event (failure event) and the number of all possible cases;

– completion of a required function: the function could be a structure supporting a constant or variable load;

– given environmental and usage conditions: these conditions are the physical, chemical, electrical and mechanical constraints submitted by a device being subjected to in its environment: ambient conditions or the actions of neighboring devices at the interface with the device;

– time: it is the time in a general sense. This will often be a number of cycles, a distance covered or, in general, an age or quantity which expresses the lifetime.

2.1.4. *Importance of reliability*

In fact, deterministic models, which feature only the averages (means) of parameters, can no longer be used for certain types of problems, as this generally leads to a highly inaccurate representation of reality. As a result, it is better to introduce uncertainties into the evaluation of parameters and consider their resulting variability. The fundamental element of these stochastic or probabilistic studies is hence to take account of the random and spatially varying nature of parameters such as material properties. The primary purpose of a reliability-based model is to determine a good confidence level of the studied structure. By making certain *a priori* hypotheses regarding the sizing parameters and random quantities, and by defining a "failure" state for the structure, we then find the evolution of the failure probability of the structure over the course of its lifetime, and verify that the sizing quantities respect the safety regulations in vigor. The application of probabilistic methods during design necessitates an effective tool for evaluating the reliability of the structures concerned. In this chapter, we present the basic principles of the reliability concepts and models: statistical measures, probability, reliability, etc.

2.2. Basic functions and concepts for reliability analysis

We present here several basic functions to better understand the reliability concept.

2.2.1. *Failure concept*

Failure is the termination in the ability of an entity to complete a required function, resulting in the breakdown of this entity. Failure criteria are defined by different categories [ELH 11]:

– by drift or degradation;

– catalectic: complete and sudden failure;

– total, partial, systematic, major, minor and catastrophic;

– software;

– human.

2.2.2. *Uncertainty concept*

Variation across the structure arises from the uncertain nature of the phenomena acting over the structure. This variability can originate from several sources: risks (or hazards) have two principal origins, intrinsic risks (operational and quality constraints) and extrinsic risks (environment constraints).

2.2.2.1. *Intrinsic risks*

Intrinsic risks which can affect the quality of a product are principally related to operational constraints (internal forces, movements, etc.) and quality constraints (material properties, surface qualities, the precision of measurements, etc.) [TEB 05]. First, the inability to perfectly model internal forces within materials is a major source of uncertainty. Second, for reasons of cost, to guarantee the same quality of raw materials can be difficult. To this effect, variability in the characteristics of the same material is observed. In order to take this variability into account, suppliers offer products with tolerance intervals. Dispersion (inhomogeneity) in material properties is represented by variability in the material characteristics. The surface quality also has a considerable effect on a structure's behavior. As a result, crack nucleation occurs more rapidly in products with poor surface treatment. Finally, human error and imprecise machining generate variations that affect the quality of the structure. These variations are manifested as geometrical inhomogeneity and cross-sectional imperfections.

2.2.2.2. *Extrinsic risks*

Extrinsic risks are primarily related to the environment and can have two origins: climatic constraints (temperature, pressure, humidity, etc.) and mechanical constraints (loading, forced deformation, etc.) [TEB 05]. First, by definition, reliability is characterized by four aspects (probability, completion of a required function, given environmental and usage conditions, and time). As a result, the structure should meet the usage conditions specified in the requirement statement. However, violations of the usage conditions can be recorded, for example loadings or temperatures exceeding the usage conditions prescribed in the requirement statement. Second, certain forces exerted on the structure are inherently uncertain, such as wind, sea swell or traffic over a bridge. These quantities vary over time and are difficult to determine precisely.

2.2.3. Random variables

A random variable is characterized by its probability distribution and the associated parameters. Modeling a random variable based on observations can be carried out in three stages:

– data collection;

– data processing (descriptive statistics);

– data interpretation, also called statistical inference, which is based upon sampling theory and mathematical statistics.

Conversely, modeling a random variable for non-observable parameters can be based upon expert opinion (information accrued with expertise).

Generally, several random variables contribute simultaneously to the modeling of a process. It is suitable to regroup random variables together into a random vector, for which we try to determine:

– a probability density function (PDF);

– a cumulative distribution function (CDF).

2.2.4. Probability density function

The PDF represents the density of the distribution over time and can be thought of as the derivative of the failure probability function (Figure 2.1). It can be written using the following form:

$$f(t) = \frac{dF(t)}{dt} = F'(t) \qquad [2.2]$$

or alternatively:

$$f(t) = \frac{dP_f(t)}{dt} = P_f'(t) \qquad [2.3]$$

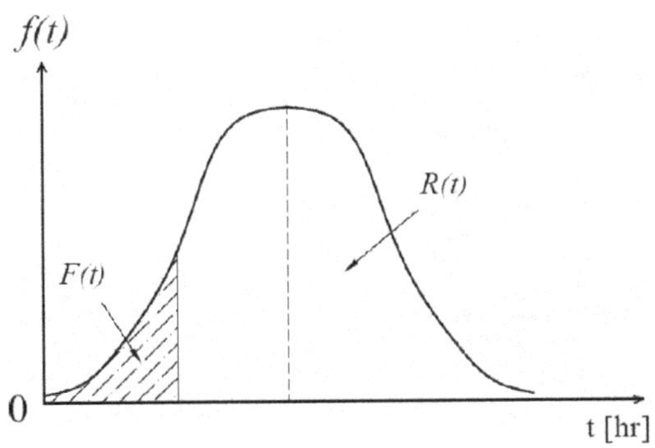

Figure 2.1. *PDF as a function of time*

2.2.5. *Cumulative distribution function*

Probability theory is a branch of mathematics providing a rigorous framework for the analysis of uncertain or random phenomena. In this chapter, we are interested in the probability of failure [CDF). A failure probability function (CDF) goes from zero up to 1 (Figure 2.2).

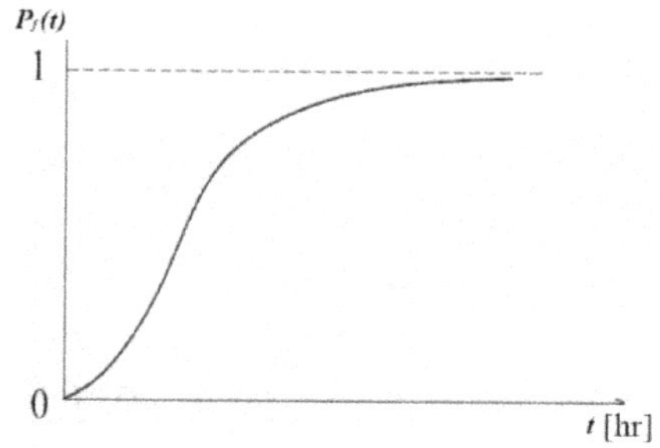

Figure 2.2. *CDF as a function of time*

2.2.6. *Reliability function*

Reliability is the complement of the probability of failure:

$$R = 1 - P_f \tag{2.4}$$

where R is the reliability and P_f is the probability of failure. In other words, reliability analysis is the same as evaluating the probability of failure. Reliability function $R(t)$ and failure probability function $F(t)$, in function of time t, are shown in Figure 2.3.

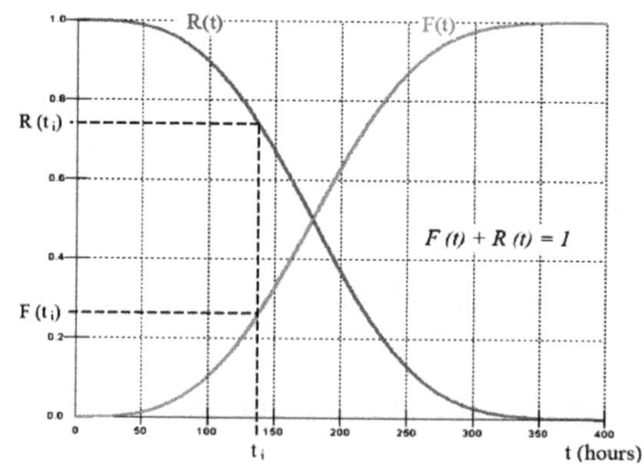

Figure 2.3. *Reliability and failure probability functions*

2.3. System reliability

A system in reliability analysis is described by a conjunction of logically components (elements). There are four types of system conjunction [KHA 11].

2.3.1. *Series conjunction*

The basic principle for a series conjunction is that failure of one component can lead to failure of the whole system. The reliability of the system can be written in the form:

$$R_s = \prod_{i=1}^{N} R_i \qquad [2.5]$$

Figure 2.4 shows an example of a series conjunction, for which a system must be made up of at least two components.

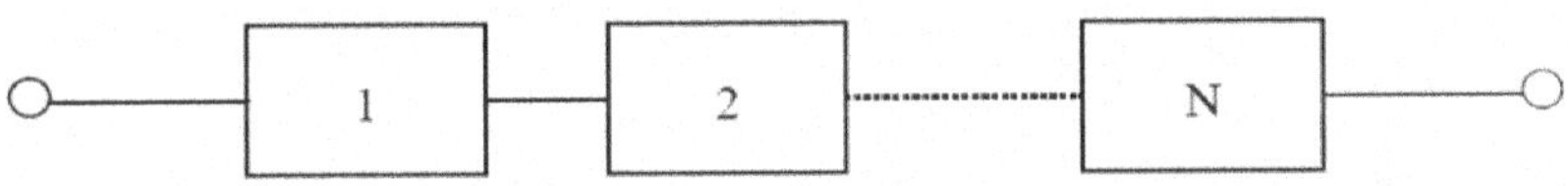

Figure 2.4. *Series configuration*

ILLUSTRATIVE EXAMPLE 2.1.–

Consider a screen card for a computer made up of 200 components, each one with a reliability ($R_i = 0.999$) of 250 working hours. To make the card work, all of its components must be working at the same time. In this case, we consider the system to be a series configuration. The system reliability value is then:

$$R_s = \prod_{i=1}^{200} 0.999 \Rightarrow R_s = (0.999)^{200} = 0.819 \qquad [2.6]$$

We note that the system reliability value is decreased when comparing to the reliability of the individual elements.

2.3.2. *Parallel conjunction*

For a parallel conjunction, the failure of one or more components does not necessarily lead to the failure of the system. In other words, a parallel system works when at least one of its components works. For example, systems used in aviation are connected in parallel, which leads to a higher level of confidence. The reliability of the system can be written in the form:

$$R_s = 1 - \prod_{i=1}^{N} (1 - R_i) \qquad [2.7]$$

Figure 2.5 shows an example of a parallel conjunction.

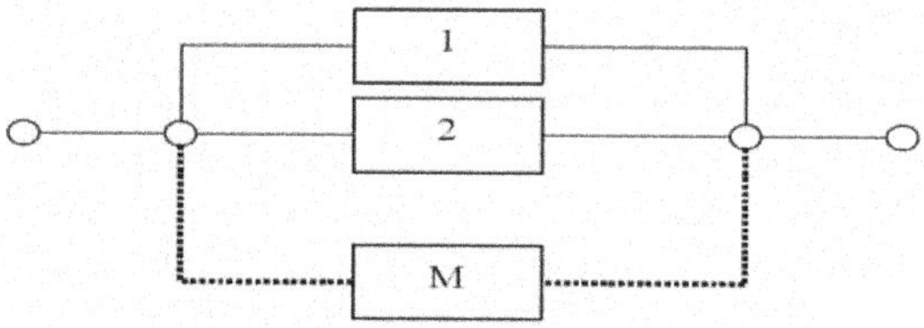

Figure 2.5. *Parallel conjunction*

ILLUSTRATIVE EXAMPLE 2.2.–

Consider screen cards, each with reliability ($R_i = 0.819$). They are connected in parallel. The system reliability (overall reliability) is then:

$$R_s = 1 - (1 - 0.819)^2 = 0.967 \qquad [2.8]$$

We note that the overall reliability is increased when comparing to the reliability of each card.

2.3.3. *Mixed conjunction*

For a mixed (series/parallel) system, the failure of one or more components belonging to the series conjunction definitely leads to the failure of the system. Figure 2.6 shows an example of a mixed conjunction.

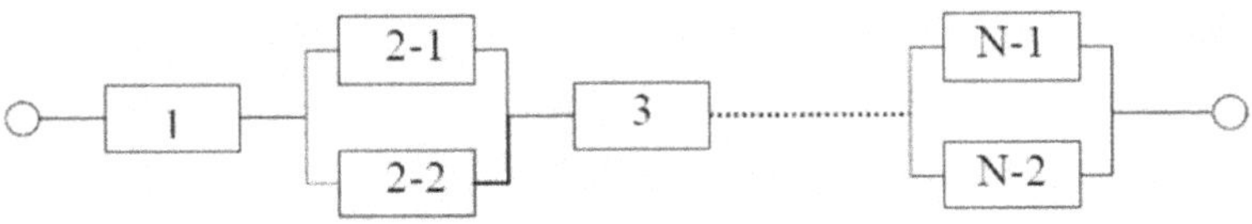

Figure 2.6. *Mixed configuration*

ILLUSTRATIVE EXAMPLE 2.3.–

Figure 2.7 shows a mixed system with two cards in parallel. Each card contains 200 components with reliability (0.998). Calculate the overall reliability of the system.

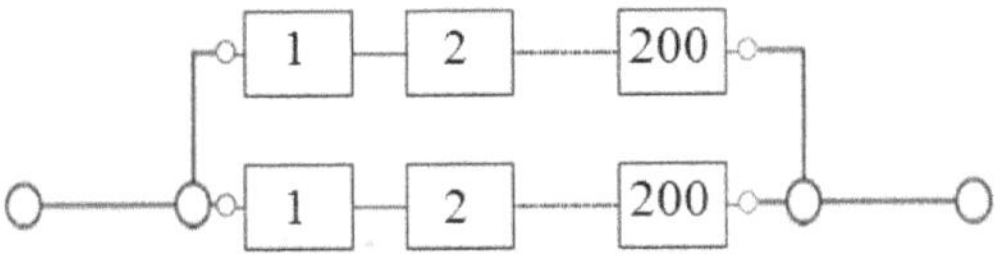

Figure 2.7. *Mixed system*

The reliability of a card (200 components in series) is then:

$$R_s = \prod_{i=1}^{200} 0.998 \Rightarrow R_s = (0.998)^{200} = 0.67 \qquad [2.9]$$

The system reliability (two cards in parallel) is then:

$$R_s = 1 - (1 - 0.67)^2 = 0.89 \qquad [2.10]$$

We note that the system reliability is greater than the reliability of each card.

2.3.4. Delta-star conjunction

To transform the delta conjunction into the star conjunction (Figure 2.8), we use the following equations:

$$R_A = \sqrt{\frac{[1-(1-R_1)\times(1-R_2\times R_3)]\times[1-(1-R_2)\times(1-R_1\times R_3)]}{1-(1-R_3)\times(1-R_1\times R_2)}} \quad [2.11]$$

$$R_B = \sqrt{\frac{[1-(1-R_1)\times(1-R_2\times R_3)]\times[1-(1-R_3)\times(1-R_1\times R_2)]}{1-(1-R_2)\times(1-R_1\times R_3)}} \quad [2.12]$$

$$R_c = \sqrt{\frac{[1-(1-R_3)\times(1-R_2\times R_1)]\times[1-(1-R_2)\times(1-R_1\times R_2)]}{1-(1-R_1)\times(1-R_3\times R_2)}} \quad [2.13]$$

When we have the special case:

$$R_1 = R_2 = R_3 = R_\Delta \quad [2.14]$$

the reliability is then:

$$R_Y = \sqrt{R_\Delta^2 + R_\Delta - R_\Delta^3} \quad [2.15]$$

Figure 2.8 shows the examples of a star conjunction and a delta conjunction.

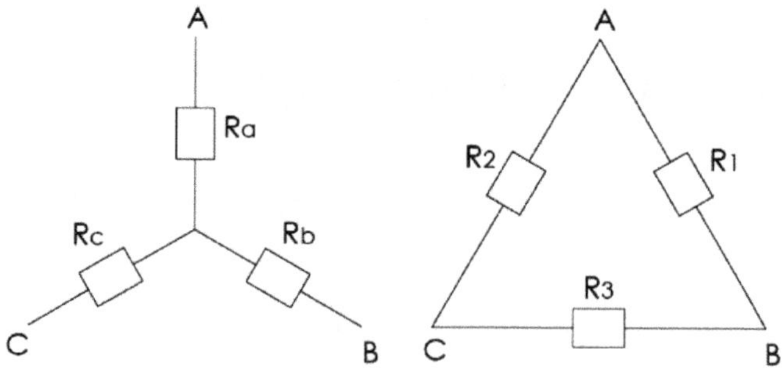

Figure 2.8. *Delta and star conjunctions*

ILLUSTRATIVE EXAMPLE 2.4.–

Consider the diagram of components presented in Figure 2.9. Calculate the reliability between the points AG, given the reliability shown for each component.

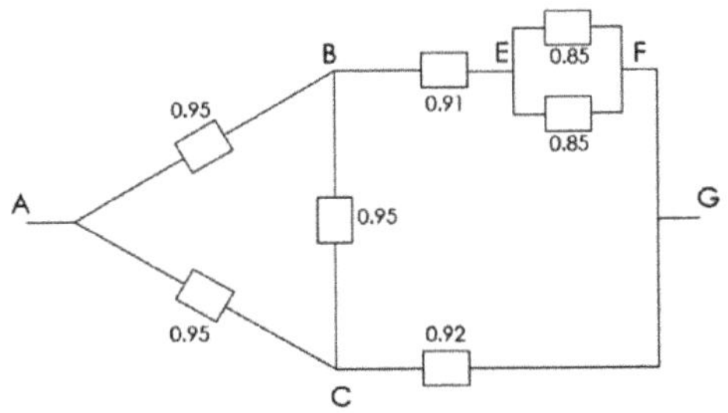

Figure 2.9. *Mixed conjunction system*

We start with a delta (ABC) to star (ABCD) transformation as follows:

$$R_Y = \sqrt{R_\Delta^2 + R_\Delta - R_\Delta^3} = \sqrt{0.95^2 + 0.95 - 0.95^3} = 0.99756 \qquad [2.16]$$

Next, we calculate the reliability between (EF) by considering a parallel conjunction using the following form:

$$R_x = 1 - \prod_{i=1}^{M} (1 - R_i) = 1 - (1 - 0.85)(1 - 0.85) = 0.9775 \qquad [2.17]$$

We thus obtain the diagram shown in Figure 2.10.

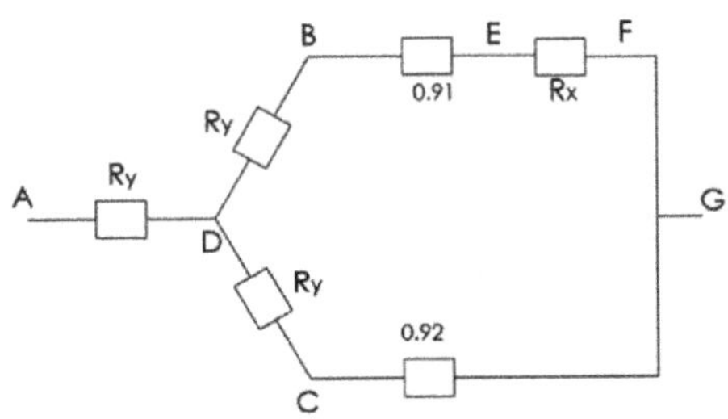

Figure 2.10. *New form of the mixed conjunction system*

Next, we calculate the reliability of the series conjunction (DBEF):

$$R_u = \prod_{i=1}^{M} R_i = R_y \times 0.91 \times R_x = (0.99756)(0.91)(0.9775) = 0.88735 \quad [2.18]$$

and the conjunction (DCG):

$$R_v = \prod_{i=1}^{M} R_i = R_y \times 0.91 = (0.99756)(0.92) = 0.917755 \quad [2.19]$$

which leads to the simplified system shown in Figure 2.11.

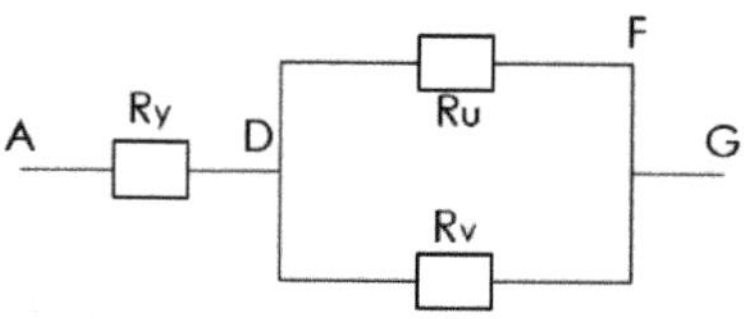

Figure 2.11. *Series conjunctions (DBEFG) and (DCG) replaced by (Ru) and (Rv), respectively*

We calculate the parallel conjunction reliability (Ru and Rv):

$$R_W = 1 - \prod_{i=1}^{M} (1 - R_i) = 1 - (1 - 0.88735)(1 - 0.917755) = 0.99074 \quad [2.20]$$

Figure 2.12 shows the result of this calculation.

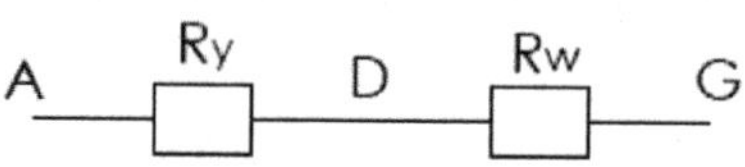

Figure 2.12. *Transformation of the parallel conjunction (RuRv) to Rw*

Finally, we calculate the reliability of the system with:

$$R_{SYS} = \prod_{i=1}^{M} R_i = R_Y \times R_W = (0.99756)(0.99074) = 0.98832$$

2.4. Statistical measures

The statistical study of random variables requires several statistical measures [KHA 11].

Expected value:

For a discrete random variable, the expected value is given by:

$$E(t_i) = \frac{\sum t_k . n_i}{\sum n_i} \qquad [2.21]$$

For a continuous random variable, the expected value is given by:

$$E(t_i) = \int_0^{\infty} t . f(t) . dt \qquad [2.22]$$

Mean value:

The mean value of a distribution of N samples is given by:

$$\mu = \frac{\sum x_i}{N} \qquad [2.23]$$

Median:

The median is the value in the middle of the samples.

Mode:

The mode is the value which has the greatest (maximum) number of samples.

Variance:

For a discrete random variable, the variance is given by:

$$V = \sum E[t_i - E(T)]^2 n_i = \sum_i t_i^2 . n_i - \left(\sum_i t_i . n_i \right)^2 \qquad [2.24]$$

For a continuous random variable, the variance is given by:

$$V = \int_0^{\infty} t^2.f(t).dt - \left(\int_0^{\infty} t.f(t).dt \right)^2 \qquad [2.25]$$

Standard deviation:

The standard deviation is given by:

$$\sigma_i = \sqrt{\frac{\sum_{k=1}^{N} (V_k - \mu)^2}{N-1}} \qquad [2.26]$$

where N is the number of samples.

Moments:

The moments are given by:

$$M_i = \sum E[t_i - E(t)]^k .n_i \qquad [2.27]$$

where k is the number of samples.

Skewness:

Skewness is given by:

$$\zeta_s = \frac{M_3}{\sigma^3} \qquad [2.28]$$

where M_3 is the third moment.

Kurtosis:

Kurtosis is given by:

$$\zeta_k = \frac{M_4}{\sigma^4} \qquad [2.29]$$

where M_4 is the fourth moment.

ILLUSTRATIVE EXAMPLE 2.5.–

Consider a set of 71 samples presented in Table 2.1. Find the different statistical measures.

18	25	42s	55	61	69	84	89	93	102	102	103
104	104	120	123	123	124	129	148	151	156	166	168
171	173	176	181	184	185	187	193	195	197	199	203
203	217	220	228	236	241	243	248	259	272	289	295
297	305	315	331	337	344	351	357	371	380	386	391
395	396	425	426	439	460	474	490	511	515	592	

Table 2.1. *A set of 71 samples*

We regroup the samples into intervals. Table 2.2 shows the different intervals and their centers, where the center of each interval is given by:

$$r = \frac{\max - \min}{1 + 3.322 \ln(N)} = \frac{603 - 18}{1 + 3.322 \ln(71)} \sum_{i=1}^{n} (X_i - \bar{X})^2$$
$$= 38.59 \approx 40 \quad [2.30]$$

The mean value can be calculated using the formula:

$$\mu = \frac{\sum n_i x_i}{\sum n_i}$$
$$= \frac{(2.20 + 4.60 + \ldots + 1.580)}{(2 + 4 + \ldots + 1)} = \frac{16900}{71} = 238.03 \quad [2.31]$$

Class	Interval	Center	Frequency	Cumulative frequency	Cumulative frequency (percentage)
1	[0–40[	20	2	2	2.8
2	[40–80[	60	4	6	8.5
3	[80–120[	100	8	14	19.7
4	[120–160[	140	8	22	31
5	[160–200[	180	13	35	49.3
6	[200–240[	220	6	41	57.7
7	[240–280[	260	5	46	64.8
8	[280–320[	300	5	51	71.8
9	[320–360[	340	5	56	78.9
10	[360–400[	380	6	62	87.3
11	[400–440[	420	3	65	91.5
12	[440–480[	460	2	67	94.4
13	[480–520[	500	3	70	98.6
14	[520–560[	540	0	70	98.6
15	[560–600[	580	1	71	100

Table 2.2. *Different intervals and their centers*

The number of intervals is 15. The median is hence the center of the eighth interval and is equal to $Me = 300$ and the mode is $Mo = 180$. The variance can be calculated using the formula:

$$V = \frac{\sum n_i (x_i - \mu)^2}{\sum n_i} = \frac{\left[2(20-238)^2 + 4(60-238)^2 + \ldots + 1(580-238)^2\right]}{71} = 17072.17 \quad [2.32]$$

The standard deviation can be calculated using the formula:

$$\sigma = \sqrt{\frac{\sum n_i (x_i - \mu)^2}{\sum n_i - 1}} = \sqrt{\frac{\left[2(20-238)^2 + 4(60-238)^2 + \ldots + 1(580-238)^2\right]}{70}} = 131.59 \quad [2.33]$$

The skewness is given by:

$$\xi_s = \frac{\dfrac{\sum n_i (x_i - \mu)^3}{\sum n_i}}{\sigma^3} = \frac{\dfrac{\left[2(20-238)^3 + 4(60-238)^3 + \ldots + 1(580-238)^3\right]}{71}}{(131.59)^3} = 0.5139 \quad [2.34]$$

The kurtosis is given by:

$$\xi_k = \frac{\dfrac{\sum n_i (x_i - \mu)^4}{\sum n_i}}{\sigma^4} = \frac{\dfrac{\left[2(20-238)^4 + 4(60-238)^4 + \ldots + 1(580-238)^4\right]}{71}}{(131.59)^4} = 2.41 \quad [2.35]$$

Figure 2.13 shows the distribution density.

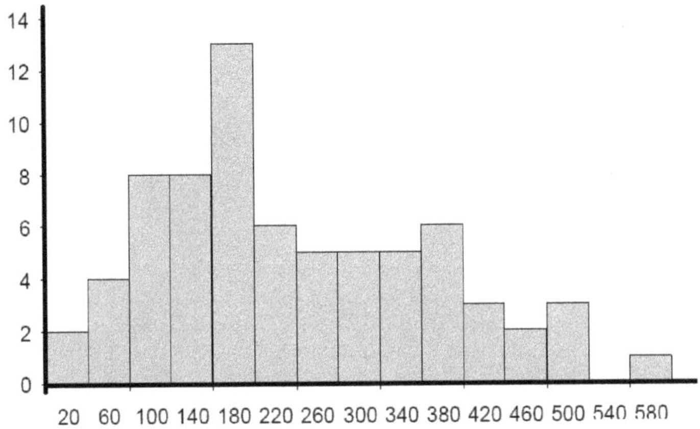

Figure 2.13. Distribution density

2.5. Probability distributions

A probability distribution is simply a list of all possible results of an experiment with their respective probabilities. Examples of the many existing distribution laws are:

– uniform distribution;

– normal distribution;

– lognormal distribution;

– Weibull distribution;

– Gumbel distribution;

– Poisson distribution;

– exponential distribution;

– β distribution;

– Pareto distribution;

– γ distribution.

Determining the probability distribution of a random variable is a challenge for engineers. Here, we present just the first three distributions (uniform, normal and lognormal laws) in detail.

2.5.1. *Uniform distribution*

The uniform probability distribution is very simple because it is the distribution of pure chance. It is a distribution of equiprobability.

2.5.1.1. *Probability density function*

The PDF is given by:

$$f(t) = \frac{1}{t_2 - t_1} \tag{2.36}$$

where the bounds of the interval in t are t_1 and t_2. Figure 2.14 shows an example of the relationship between the probability density function and time for the uniform probability distribution.

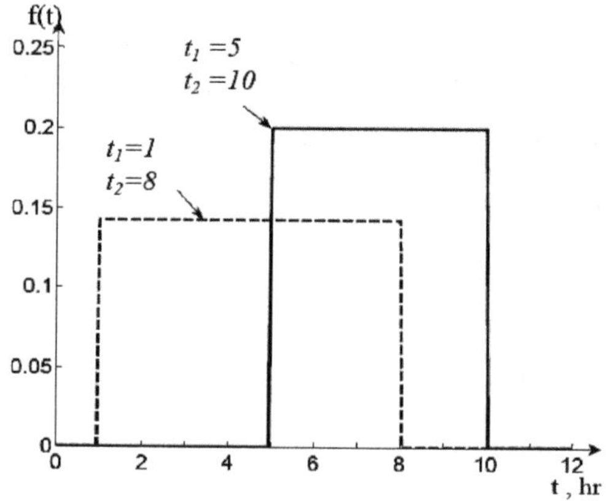

Figure 2.14. *Probability density function for the uniform distribution*

2.5.1.2. *Cumulative distribution function*

The CDF is given by:

$$F(t)=\frac{t-t_1}{t_2-t_1} \quad [2.37]$$

Figure 2.15 shows an example of the relationship between the CDF and time for the uniform probability distribution.

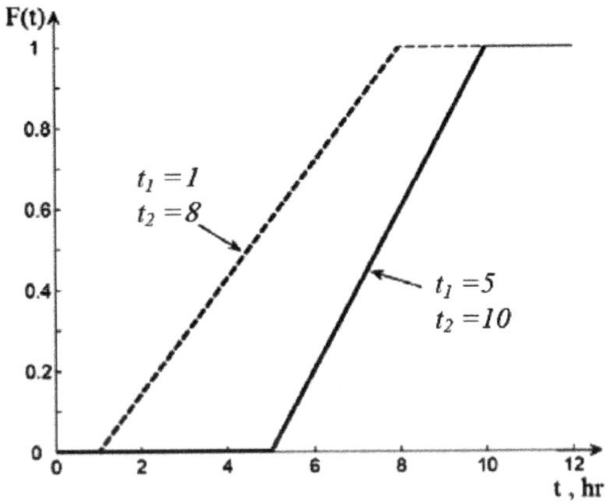

Figure 2.15. *Cumulative distribution function for the uniform probability distribution*

2.5.1.3. *Reliability function*

The reliability function is given by:

$$R(t)=\frac{t_2-t}{t_2-t_1} \quad [2.38]$$

The mean value is given by:

$$\mu=\frac{t_1+t_2}{2} \quad [2.39]$$

The variance is given by:

$$V = \frac{(t_2 - t_1)^2}{12} \quad [2.40]$$

The median is given by:

$$Me = (t_2 - t_1)/2 \quad [2.41]$$

Figure 2.16 shows an example of the relationship between the reliability function and time for the uniform probability distribution.

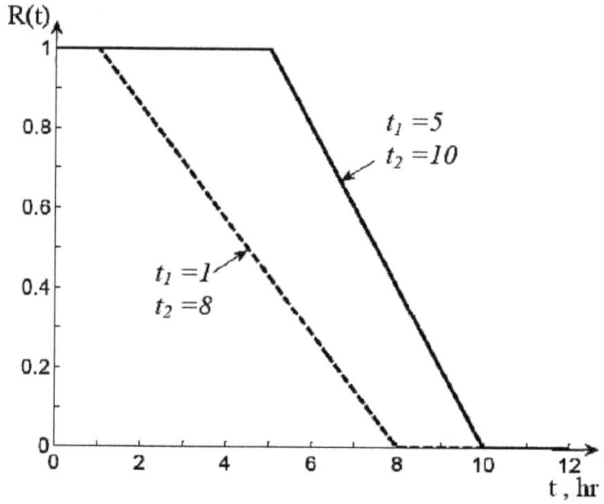

Figure 2.16. *Reliability function for the uniform probability distribution*

ILLUSTRATIVE EXAMPLE 2.6.–

Consider a set of 21 samples presented in Table 2.3. Find the failure probability for a pen after 35 days.

5	14	16	22	26	36	38	43	46	48	57
60	63	70	74	75	80	87	92	95	97	

Table 2.3. *A set of 21 values*

Table 2.4 shows the different intervals and their centers.

Interval	Frequency	Cumulative frequency	Cumulative frequency (percentage)
[0–10]	1	1	4.8
[10–20]	2	3	14.3
[20–30]	2	5	23.8
[30–40]	2	7	33.3
[40–50]	3	10	47.6
[50–60]	1	11	52.4
[60–70]	2	13	61.9
[70–80]	3	16	76.2
[80–90]	2	18	85.7
[90–100]	3	21	100

Table 2.4. *Different intervals and their frequency*

Modeling this set leads to a uniform probability distribution as shown in Figure 2.17.

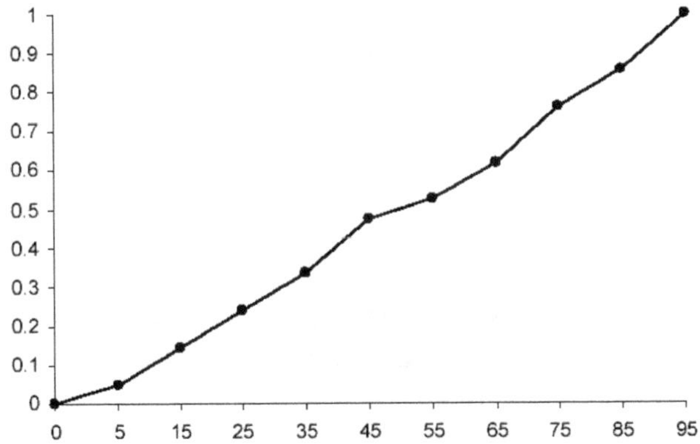

Figure 2.17. *Cumulative density function for the given data*

We note that the curve starts at $t_1 = 0$ and stops at $t_2 = 95$. Using equation [2.36], the PDF can be calculated as follows:

$$f(t) = \frac{1}{95} = 0.0105 \qquad [2.42]$$

Using equation [2.37], the failure probability function (CDF) can be calculated in the form:

$$F(t) = \frac{t}{95} = 0.0105 \times t \qquad [2.43]$$

The reliability function can be calculated in the form:

$$R(t) = 1 - 0.0105 \times t \qquad [2.44]$$

At 35 days, the failure probability value is equal to $P_f(35) = 0.3684$ (equation [2.43]).

2.5.2. *Normal distribution*

The normal (or Gaussian) distribution is one of the most common probability distributions because it applies to numerous phenomena. It is characterized by the following.

2.5.2.1. *Probability density function*

The PDF is given by:

$$f(t) = \frac{1}{\sigma.\sqrt{2\pi}}.\exp\left\{-\frac{(t-\mu)^2}{2\sigma^2}\right\} \qquad [2.45]$$

Figure 2.18 shows an example of the relationship between the probability density function and time for the normal distribution.

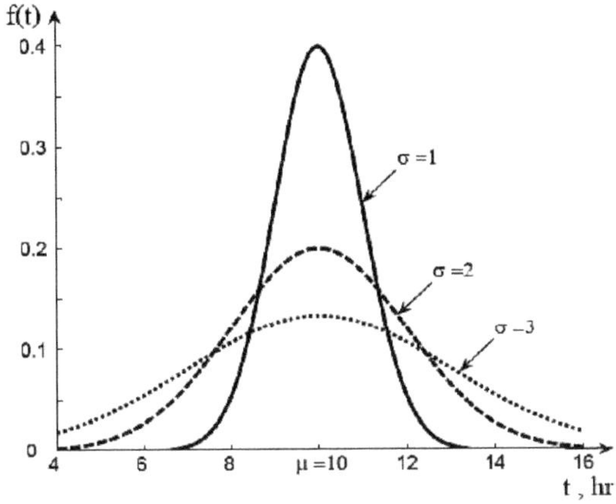

Figure 2.18. *Probability density function (PDF) for the normal distribution*

2.5.2.2. Cumulative distribution function

The CDF is given by:

$$F(t) = \frac{1}{\sigma.\sqrt{2\pi}} \int_{-\infty}^{t} \exp\left\{ -\frac{(t-\mu)^2}{2\sigma^2} \right\}.dt \qquad [2.46]$$

Figure 2.19 shows an example of the relationship between the CDF and time for the normal distribution.

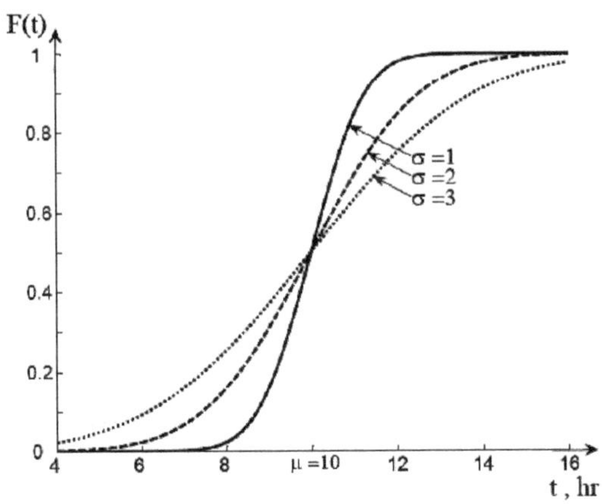

Figure 2.19. *Cumulative distribution function (CDF) for the normal distribution*

2.5.2.3. *Reliability function*

The reliability function is given by:

$$R(t) = \frac{1}{\sigma.\sqrt{2\pi}} \int_{t}^{+\infty} \exp\left\{ -\frac{(t-\mu)^2}{2\sigma^2} \right\}.dt \qquad [2.47]$$

The mean value is given by:

$$\mu = \frac{\sum_{i=1}^{n} t_i}{n} \qquad [2.48]$$

The variance is given by:

$$V = \sqrt{\frac{\Sigma(t_i - \mu)^2}{n}} \qquad [2.49]$$

The standard deviation is given by:

$$\sigma = \sqrt{\frac{\Sigma(t_i - \mu)^2}{n-1}} \qquad [2.50]$$

The median and the mode are equal to the mean value ($Me = Mo = \mu$). Figure 2.20 shows an example of the relationship between the reliability function and time for the normal distribution.

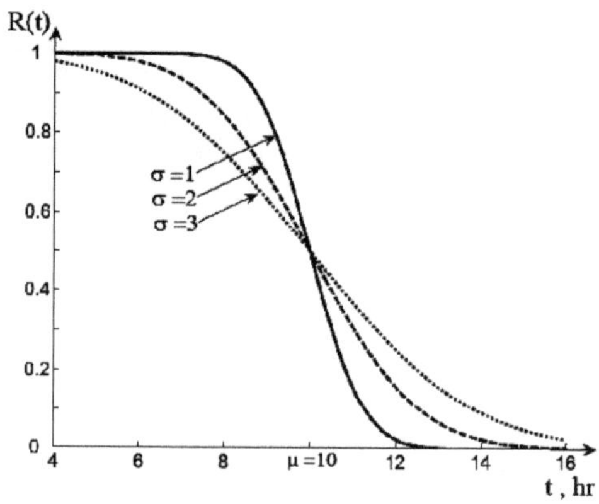

Figure 2.20. *Reliability function for the normal distribution*

ILLUSTRATIVE EXAMPLE 2.7.–

Consider a set of data representing 46 ball-bearings. The values presented in Table 2.5 are the number of journeys (in millions) after which the ball-bearing failed:

– calculate the mean value, standard deviation and normalized value for the random variable when its absolute value equals 100, given an interval of width 30;

– calculate the failure probability and reliability values when the random variable is equal to 100.

9	110	120	125	133	142	150	152	159	161
164	164	176	176	180	181	188	194	195	199
200	205	207	207	208	212	215	217	218	218
220	224	228	238	249	252	252	255	257	259
265	277	286	289	318	351				

Table 2.5. *A set of 46 samples*

We divide the intervals by their width, 30, such that the first interval is $[90-120[$. Table 2.6 shows the different intervals and their centers.

Interval	Center	Frequency	Cumulative frequency	Cumulative frequency (percentage)
[90–120[	105	2	2	4.3
[120–150[	135	4	6	13.0
[150–180[	165	8	14	30.4
[180–210[	195	11	25	54.3
[210–240{	225	9	34	73.9
[240–270[	255	7	41	89.1
[270–300[	285	3	44	95.7
[300–330[	315	1	45	97.8
[330–360[	345	1	46	100.0

Table 2.6. *Different intervals and their centers*

When modeling the distribution curve, we find that the distribution resembles the normal distribution (Figure 2.21).

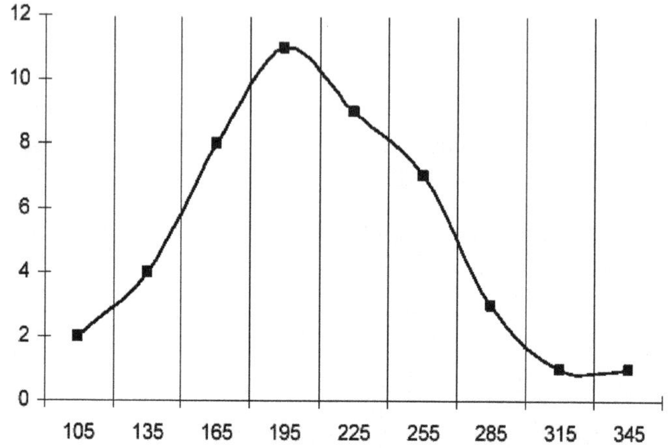

Figure 2.21. *Normal distribution of values*

The mean value can be calculated as follows:

$$\mu = \frac{105 \times 2 + ... + 345 \times 1}{46} = 207.39 \quad [2.51]$$

The standard deviation can be calculated as follows:

$$\sigma = \sqrt{\frac{2 \times (105 - 207.39)^2 + ... + 1 \times (345 - 207.39)^2}{46 - 1}} = 52.75 \quad [2.52]$$

Using equation [2.45], the PDF can be calculated as follows:

$$f(t) = \frac{1}{52.75.\sqrt{2\pi}}.\exp\left\{-\frac{(t-207.4)^2}{2(52.75)^2}\right\} = \frac{1}{132.2} e^{\frac{(t-207.4)^2}{-5566.1}} \quad [2.53]$$

Using equation [2.46], the failure probability function (CDF) can be calculated as follows:

$$F(t) = \frac{1}{52.75.\sqrt{2\pi}} \int_{-\infty}^{t} \exp\left\{-\frac{(t-207.4)^2}{2(52.75)^2}\right\}.dt = \int_{-\infty}^{t} \frac{1}{132.2} e^{\frac{(t-207.4)^2}{-5566.1}} dt \qquad [2.54]$$

The reliability can be calculated in the form:

$$R(t) = \frac{1}{52.75.\sqrt{2\pi}} \int_{t}^{+\infty} \exp\left\{-\frac{(t-207.4)^2}{2(52.75)^2}\right\}.dt$$
$$= \frac{1}{132.2} \int_{t}^{\infty} e^{\frac{(t-207.4)^2}{-5566.1}} dt \qquad [2.55]$$

When the random variable value is t=100, the numerical integral [2.54] leads to a failure probability value equal to:

$$P_f(100) = \int_{-\infty}^{100} \frac{1}{132.2} e^{\frac{(100-207.4)^2}{-5566.1}} dt \approx 0.02 \qquad [2.56]$$

Using equation [2.4], the reliability value is then:

$$R(100) = 1 - P_f(100) \approx 0.98 \qquad [2.57]$$

2.5.3. Lognormal distribution

A continuous and positive random variable, t, follows a lognormal distribution if its logarithm is normally distributed. This distribution is used in reliability analysis to model failures caused by fatigue. It is characterized by the following.

2.5.3.1. *Probability density function*

The PDF is given by:

$$f(t) = \frac{1}{\sigma . t \sqrt{2\pi}} . \exp\left\{ -\frac{(\ln t - \mu)^2}{2\sigma^2} \right\} \qquad [2.58]$$

Figure 2.22 shows an example of the relationship between the density function and time for a lognormal distribution.

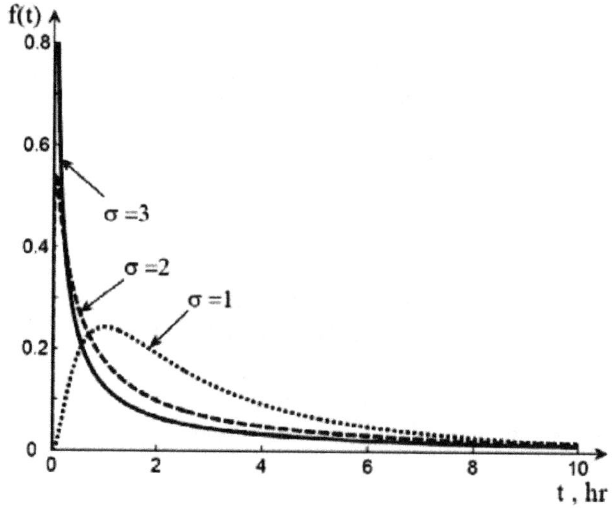

Figure 2.22. *Probability density function for a lognormal distribution*

2.5.3.2. *Cumulative distribution function*

The CDF is given by:

$$F(t) = \frac{1}{\sigma . \sqrt{2\pi}} \int_{-\infty}^{t} \frac{1}{t} \exp\left\{ -\frac{(\ln t - \mu)^2}{2\sigma^2} \right\} .dt \qquad [2.59]$$

Figure 2.23 shows an example of the relationship between the CDF and time for a lognormal distribution.

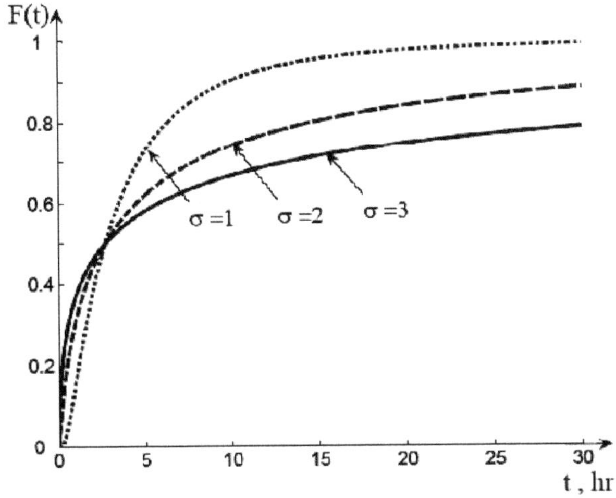

Figure 2.23. *Cumulative distribution function for a lognormal distribution*

2.5.3.3. *Reliability function*

The reliability function is given by:

$$R(t) = \frac{1}{\sigma.\sqrt{2\pi}} \int_{t}^{+\infty} \frac{1}{t} \exp\left\{ -\frac{(\ln t - \mu)^2}{2\sigma^2} \right\}.dt \qquad [2.60]$$

The mean value is given by:

$$\mu = \frac{\sum \ln t_i}{N} \qquad [2.61]$$

The variance is given by:

$$V = e^{2\mu+\sigma^2}(1 - e^{\sigma^2}) \qquad [2.62]$$

The standard deviation is given by:

$$\sigma = \sqrt{\frac{\sum (\ln t_i - \mu)^2}{N-1}} \qquad [2.63]$$

The mode is given by:

$$Mo = e^{\mu - \sigma^2} \qquad [2.64]$$

The median is given by:

$$Me = e^{\mu} \qquad [2.65]$$

The expected value is given by:

$$E(T) = e^{\mu + \frac{\sigma^2}{2}} \qquad [2.66]$$

Figure 2.24 shows an example of the relationship between the reliability function and time for a lognormal distribution.

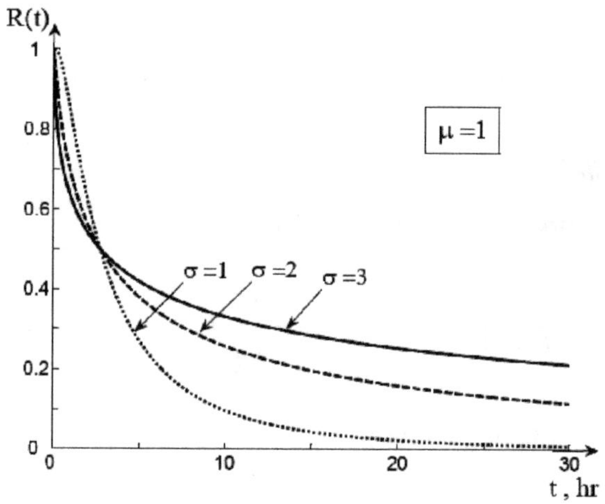

Figure 2.24. *Reliability function for a lognormal distribution*

ILLUSTRATIVE EXAMPLE 2.8.–

Consider a set of data representing 62 camshafts. The samples presented in Table 2.7 are the number of journeys (in millions) after which the bearing fails:

– calculate the mean value, the standard deviation and the normalized value for the random variable when its absolute value equals 20, given the width of the interval equals 60;

– calculate the reliability functions.

12	12	23	37	38	39	48	49	49	50	51	51
53	55	61	65	66	72	75	79	92	95	102	104
109	109	109	118	121	123	135	151	152	156	157	158
167	173	179	190	200	200	202	202	203	206	210	241
247	265	271	283	293	328	374	376	384	486	517	578
641	784										

Table 2.7. *A set of 62 samples*

We separate into intervals of width 60 such that the first interval is $[10-70[$. Table 2.8 shows the different intervals and their centers.

Intervals	Centers	Log(centers)	Frequency	Cumulative frequency	Cumulative frequency (percentage)
[10–70]	40	3.69	17	17	27.4
[70–130]	100	4.61	13	30	48.4
[130–190]	160	5.08	9	39	62.9
[190–250]	220	5.39	10	49	79.0
[250–310]	280	5.63	4	53	85.5
[310–370]	340	5.83	1	54	87.1
[370–430]	400	5.99	3	57	91.9
[430–490]	460	6.13	1	58	93.5
[490–550]	520	6.25	1	59	95.2
[550–610]	580	6.36	1	60	96.8
[610–670]	640	6.46	1	61	98.4
[670–730]	700	6.55	0	61	98.4
[730–790]	760	6.63	1	62	100.0

Table 2.8. *The different intervals and their centers*

When modeling the distribution curve, we find that the distribution resembles the lognormal distribution (Figure 2.25).

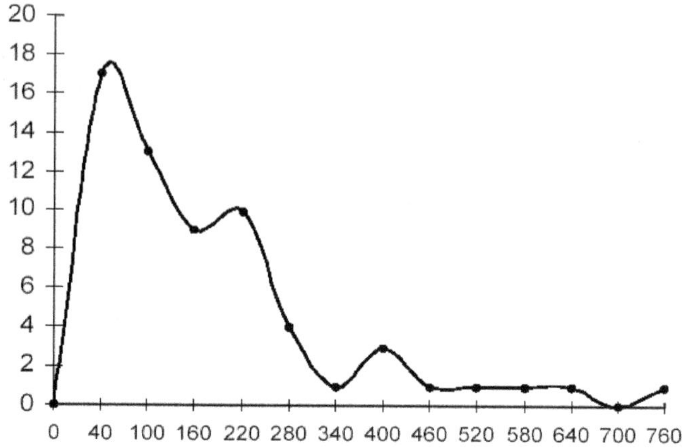

Figure 2.25. *Lognormal distribution of values*

The mean value can be calculated as follows:

$$\mu = \frac{17 \times \ln(40) + ... + 1 \times \ln(760)}{62} = 4.845 \quad [2.67]$$

The standard deviation can be calculated as follows:

$$\sigma_i = \sqrt{\frac{17 \times [\ln(40) - 4.845]^2 + ... + 1 \times [\ln(760) - 4.845]^2}{62 - 1}} = 0.8695 \quad [2.68]$$

Using equation [2.58], the PDF can be calculated as follows:

$$f(t) = \frac{1}{0.869 \times t \times \sqrt{2\pi}} . \exp\left\{ -\frac{(\ln t - 4.845)^2}{2(0.869)^2} \right\} = \frac{1}{2.178 \times t} e^{\frac{(\ln t - 4.845)^2}{-1.51}} \quad [2.69]$$

Using equation [2.59], the failure probability function (CDF) can be calculated as follows:

$$F(t) = \frac{1}{0.869 \times \sqrt{2\pi}} \int_{-\infty}^{t} \frac{1}{t} \exp\left\{ -\frac{(\ln t - 4.845)^2}{2(0.869)^2} \right\} .dt = \frac{1}{2.178} \int_{-\infty}^{t} \frac{1}{t} e^{\frac{(\ln t - 4.845)^2}{-1.51}} dt \quad [2.70]$$

The reliability can be calculated as follows:

$$R(t) = \frac{1}{0.869\times\sqrt{2\pi}} \int_{t}^{+\infty} \frac{1}{t}\exp\left\{-\frac{(\ln t - 4.845)^2}{2(0.869)^2}\right\}.dt = \frac{1}{2.178}\int_{t}^{\infty} \frac{1}{t} e^{\frac{(\ln t - 4.845)^2}{-1.51}} dt \quad [2.71]$$

When the random variable value is $t = 20$, the numerical integral [2.71] leads to a failure probability value equal to:

$$P_f(20) = \frac{1}{2.178}\int_{-\infty}^{20} \frac{1}{20} e^{\frac{(\ln 20 - 4.845)^2}{-1.51}} dt \approx 0.017 \quad [2.72]$$

Using equation [2.4], the reliability value is then:

$$P_f \approx 0.017 \Rightarrow R(20) \approx 0.983 \quad [2.73]$$

The numerical integrals require a high computing time. In the next section, we present the reliability analysis strategy to calculate simply the probability of failure.

2.6. Reliability analysis

Reliability analysis approaches are techniques used as a tool to help take account of all possible uncertainties during design and construction of a structure in order to calculate the probability of failure (or reliability level) corresponding to one or more failure scenarios.

2.6.1. *Definitions*

When performing reliability analysis for structures, the governing parameters are taken to be random variables with certain probability distributions, which we use in equations describing the physics of a failure to calculate the probability of failure. The objective is thus to calculate the probability of a structure suffering failure. Here, we first limit ourselves to the reliability of one component. We then define random variables versus deterministic variables.

2.6.1.1. *Random variables versus deterministic variables*

We distinguish between two types of variables: (1) *optimization variables* **x** are deterministic variables to be controlled in order to optimize the structure specifications; they represent the control parameters of the mechanical system (i.e. dimensions, materials, loads, etc.) and of the probabilistic model (i.e. mean values and standard deviations of the random variables). In order to define the domain of feasible solutions, optimization constraints are introduced in the space x and are denoted by $g_k(\mathbf{x})$; (2) *random variables* **y** represent the uncertainties in the structural system. Each of these variables is identified by its probability distribution and the associated parameters. These variables could be geometric dimensions, material characteristics or exterior loads.

2.6.1.2. *Probability of failure*

Although rare, events such as bursting pipes, reservoirs or enclosures can have extremely serious consequences. Until recently, we protected ourselves from these incidents by using safety margins (safety factors) accumulating with each governing parameter, using a purely deterministic methodology. However, this approach does not allow a thorough understanding of the risks being taken. The probability being calculated is that of the limit state function *G*(.) returning a negative value, this is denoted by P_f (probability of failure). We first consider a random variable vector denoted by **y**. In order to calculate the probability of failure, a multidimensional integral must be evaluated:

$$P_f = \int_{G(\mathbf{y})\le 0} f(\mathbf{y})\, dy_1 \cdots dy_n \qquad [2.74]$$

where P_f is a common function calculated from the PDF of a random vector **y**. Figure 2.26 shows three examples of probability distributions (normal, lognormal and uniform distributions). It is very difficult to evaluate the integral for the probability of failure using numerical integration as shown in the previous section. Monte-Carlo simulation is a simple method which uses the fact that the integral for the probability of failure can be interpreted as the mean value of a stochastic experiment. A value is thus given by taking the average of a suitably large number of independent results (simulations) of this experiment. This method is not calculation time-efficient for probabilities greater than 10^{-6} [JAL 06]. FORM/SORM reliability methods use a combination of analytical and approximation methods to estimate the probability of failure [KAD 07].

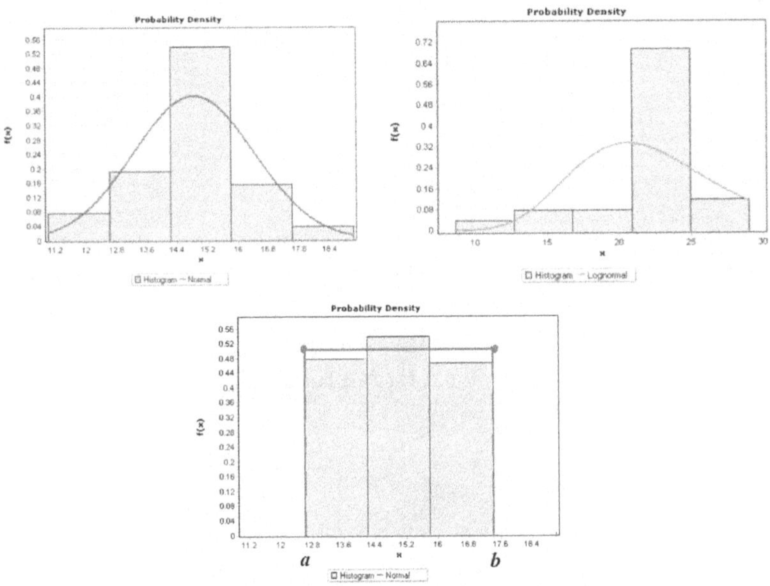

Figure 2.26. *Statistical study of probability density functions*

2.6.1.3. *Limit state function*

When applying an action $A(.)$ to a structure of a given state $K(.)$ and resistance $R(.)$, we obtain a solicitation or response $S(.)$. This is also called the performance function and is denoted by $G(.)$, its intersection with the hyper-plan of variables defines the border between the safe region and the failure region. A positive value of G signifies that the realizations of random variables lead to a safe outcome, while a negative value signifies that they lead to a failure outcome. The curve (or surface) represented by $G(.)=0$ is called the limit state curve (or surface). When the state function is written in the form of the difference between a resistance R and a solicitation S, $G = R - S$, it is called the margin of safety.

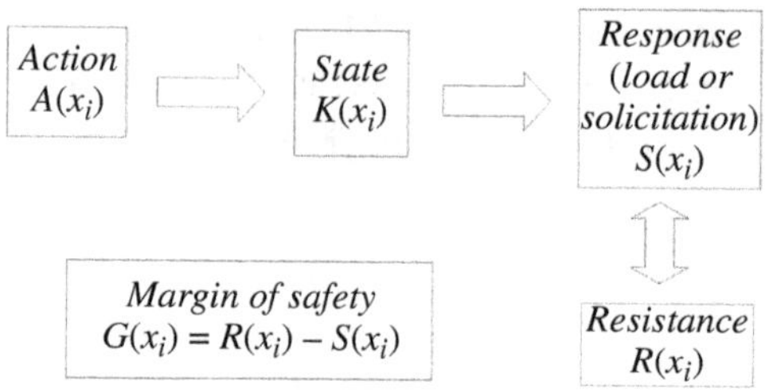

Figure 2.27. *Summary of the margin of safety*

Figure 2.28 shows a comparison between two probabilities of failure scenarios. The probability of failure P_f can be seen as the difference between the PDF for resistance and the PDF for solicitation. Figure 2.28(a) shows a small probability of failure, whereas Figure 2.28(b) shows a large probability of failure.

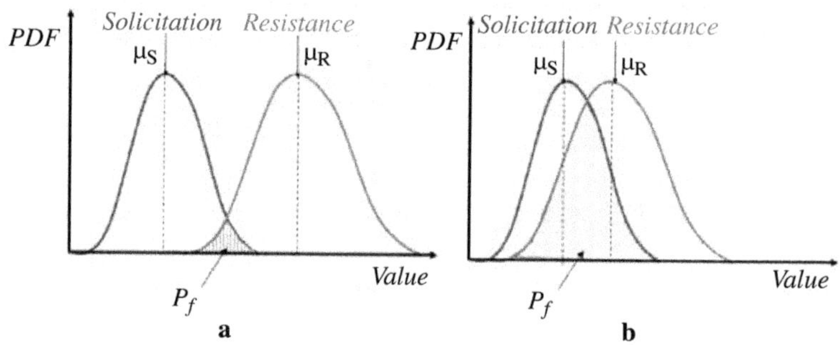

Figure 2.28. *Probabilities of failure*

ILLUSTRATIVE EXAMPLE 2.9.–

To understand the different definitions of reliability, let us consider the specifications for a shaft subjected to a normal force by the application of traction/compression loading [LEM 09]. The mechanical solicitation in this case is a stress of magnitude P/A.

The specification requirement is that the stress must not exceed the yield stress value. This can be expressed as:

$$\frac{P}{A} < \sigma_y \qquad [2.75]$$

The vector of basis variables is $\mathbf{y} = \{P, A, \sigma_y\}$, hence the limit state function is expressed as $G(\mathbf{y}) = \sigma_y A - P$ and the quantity $\sigma_y A$ is the resistance, denoted by *R*, and *P* is the solicitation, denoted by *S*.

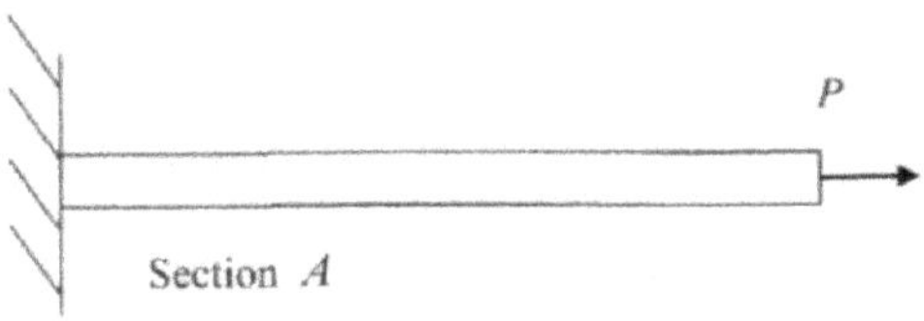

Figure 2.29. *Shaft subject to a normal force*

For a simple representation of the probability of failure, we assume that only two variables are random, *P* and σ_y. The vector of basis variables is $\mathbf{y} = \{P, \sigma_y\}^t$, the limit state function is thus expressed as: $G(\mathbf{y}) = \sigma_y A - P$ and the quantity $\sigma_y A$ is the resistance, denoted by *R*, and *P* is the solicitation, denoted by S. The probability of failure can be written as:

$$P_f = \int_{\sigma_y A - P \leq 0} f_{P,A,\sigma_y}(\sigma_y A, P) A \, d\sigma_y \, dP \qquad [2.76]$$

2.6.1.4. *Design point*

This is also called the most probable failure point (MPP) or critical point. It is the point in the failure zone which has the highest probability of occurring when determining values for the random variables. It is the point on the border between the safe and failure domains that is closest to the origin in the space of the reduced and centered variables.

ILLUSTRATIVE EXAMPLE 2.10.–

Figure 2.30 shows a modeling of a problem of a shaft under tension. The mechanical solicitation in this case is S with a PDF $f(S)$ and resistance denoted by R with a PDF $f(R)$. The limit state surface $G(R,S)=0$ separates space into two regions: a safe domain ($R>S$) and a failure domain ($R<S$). The design point is situated at the point on the limit state surface $G(R,S)=0$ that is nearest to the mean value (μ_R,μ_S).

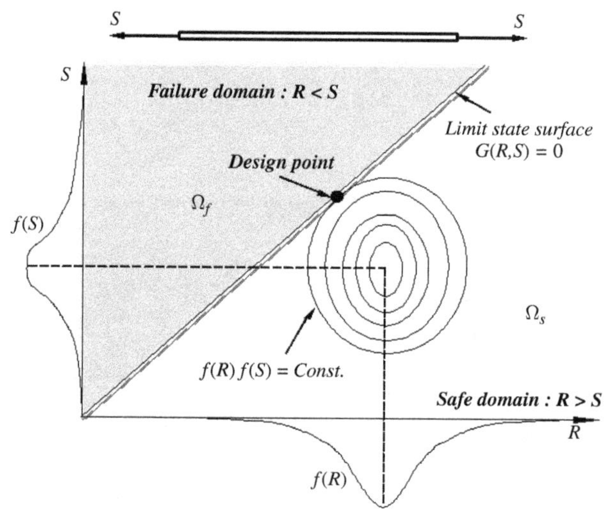

Figure 2.30. *Geometrical modeling of the design point*

2.6.1.5. *Reliability index*

The reliability level is described by a reliability index. The reliability index is hence a measure of the system safety. It is directly related to the probability of failure and allows the comparison of different systems. The limit state function plays a major role in algorithms evaluating the reliability index, hence the importance for it to have "good" regularity properties (continuity and differentiability). We are sometimes led to carry out a change of variables, in which case the mathematical expression of the limit state function must be reformulated in the new space. There are many different types of reliability indices in the literature, among which we can mention the Cornell reliability index and the Hasofer–Lind reliability index.

2.6.1.5.1. Cornell reliability index

The Cornell index is defined [COR 67] in terms of the margin of safety M, according to the mean value and standard deviation by the following equation:

$$\beta_C = \frac{E(M)}{\sqrt{V(M)}} \qquad [2.77]$$

In the one-dimensional case, the fracture surface is represented by the point M=0. The idea expressed by this definition of the reliability index is that the distance from the mean value E(M) to the fracture surface (expressed as a number of standard deviations) gives a good measure of the reliability. If the limit state function is linear, then E(M) and V(M) can easily be calculated as a function of the basis variable characteristics. In the case that the limit state is not linear, Cornell suggests a first-order Taylor approximation at the mean point. This evaluation only requires knowledge of the first two moments of the random variables. We can equally resort to Monte-Carlo simulations for complex or implicit forms of the limit state: we generally obtain good estimations of E(M) and V(M) with a relatively limited number of simulations.

2.6.1.5.2. Hasofer–Lind reliability index

The most commonly used index, denoted by β, was proposed by Hasofer and Lind [HAS74]. These authors proposed that the problem be considered not in the space of physical variables, but rather to perform a change of variables and then work in a space of statistically independent reduced Gaussian variables (mean values to be zero and standard deviations to be units) [KAD07]. The reliability is expressed by the probability of success in a scenario represented by a performance function $H(\mathbf{u})$ (this function is chosen such that $H(\mathbf{u}) \leq 0$ signifies the failure outcome). A reliability index β is introduced as a measure of the level of reliability. In a normalized space, the reliability index β is calculated by an optimization procedure constrained to the failure domain:

$$\begin{array}{ll} \min\limits_{\mathbf{u}} & d(\mathbf{u})=\sqrt{\sum u_i^2} \\ \text{subject to} & H(\mathbf{u})=0 \end{array} \qquad [2.78]$$

where $\mathbf{u} = T(\mathbf{y})$ is the iso-probabilistic transformation between the physical and normalized spaces (Figure 2.31); the physical variables $\mathbf{y}$ are calculated by the inverse transformation: $\mathbf{y} = T^{-1}(\mathbf{u})$.

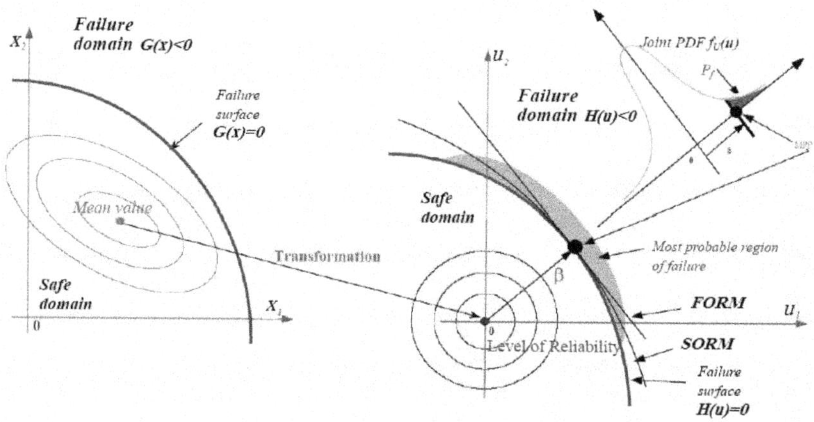

Figure 2.31. *Transformation between the physical space and the normalized space [KHA 08]*

Solving equation [2.78] poses classical optimization problems: the existence of local minima, gradient approximations and calculation times.

ILLUSTRATIVE EXAMPLE 2.11.–

Consider a limit state function $G(R,S) = 6R - 4S$. We assume that the variables R and S are not correlated and have mean values of 5 and 6, respectively. The standard deviations are 0.2 and 0.4, respectively. Calculate the Cornell and Hasofer–Lind reliability indices.

For the Cornell index, we calculate the mean value and standard deviation as follows:

$$m_G = 6 \times 5 - 4 \times 5 = 6 \quad [2.79]$$

$$\sigma_G = \sqrt{6^2 \times 0.2^2 + 4^2 \times 0.4^2} = 2 \quad [2.80]$$

The Cornell reliability index is then:

$$\beta_C = \frac{m_G}{\sigma_G} = \frac{6}{2} = 3 \quad [2.81]$$

To calculate the Hasofer–Lind reliability index, we perform a variable transformation to the normalized space: $y = m_y + u_y \sigma_y$. The limit state becomes:

$$H = 6 \times (m_R + u_R \sigma_R) - 4 \times (m_S + u_S \sigma_S) \quad [2.82]$$

$$H = 3 + 0.6 u_R - 0.8 u_S \quad [2.83]$$

The Hasofer–Lind reliability index is:

$$H = \beta_{HL} + \alpha_R u_R + \alpha_S u_S = 0 \quad [2.84]$$

from which $\beta_{HL} = 3$, $\alpha_R = 0.6$ and $\alpha_S = -0.8$

2.6.2. *Algorithms*

The first stage in reliability analysis consists of defining the random variables **y** presenting a significant level of fluctuation. These variables, called the basis variables, can be exterior actions (loads, wind, sea swell and earthquakes), geometrical characteristics (dimensions, areas and moments of inertia) or material properties (yield stress, Young's modulus and Poisson's ratio). We assign each of these variables a probability distribution describing the corresponding risk. This can be obtained through statistical studies, physical observations or, in the absence of other methods, expert opinion. The quality of this information is reflected in the precision of the results obtained. The second stage consists of defining a certain number of possible failure scenarios. A performance function $H(\mathbf{u})$ is established for each of these (for example, the resistance being greater than the solicitation, or a displacement being smaller than an acceptable value). This way, the performance function divides the space of variables into two regions: a safe domain $H(\mathbf{u})>0$ and a failure domain $H(\mathbf{u})<0$. The boundary between these domains is defined by $H(\mathbf{u})=0$, called the *limit state*. The algorithm consists of a simple single loop containing an evaluation of objective and limit state functions. Next, a simple limit state test is carried out to provide results or to update the normalized vector until convergence (Figure 2.32).

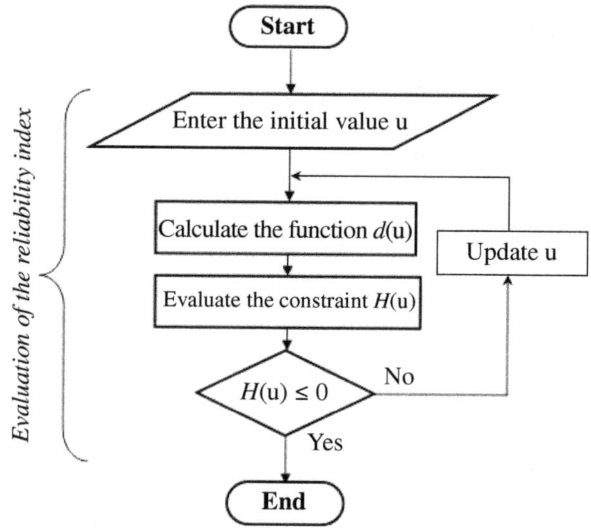

Figure 2.32. *Algorithm for evaluating the reliability index*

2.6.3. *Reliability analysis methods*

Reliability analysis is carried out using several methods based upon numerical integration such as the Monte-Carlo simulation method, approximations like the FORM and SORM methods, or response surfaces.

2.6.3.1. *Monte-Carlo simulation method*

The principle of the Monte-Carlo method presented in Figure 2.33 is based upon the generation of a large number of random draws. For each draw, the performance function is evaluated to see if the drawn configuration is located in the safe or failure domains. The probability of failure is then simply the number of failure configurations over the total number of draws.

The Monte-Carlo method is very robust and constitutes the most reliable method of evaluating the probability of failure. However, this method is very costly in terms of computing time, especially for real systems with low probabilities of failure, since for a probability of failure of 10^{-n}, between 10^{n+2} and 10^{n+3} draws must be made.

Several variants of this method exist, each aiming to optimize the strategy in an effort to reduce the calculation cost, which is essentially related to the number of calls of finite element models.

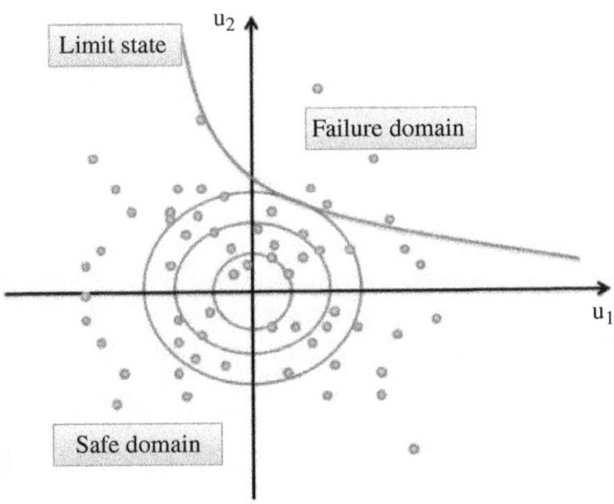

Figure 2.33. *Monte-Carlo simulation in standard space*

2.6.3.2. *Approximation methods*

The best-known methods are the first- and second-order FORM/SORM methods (first-order reliability method/second-order reliability method), which are very effective for calculating reliability. These methods are based on searching for the design point, also called the MPP, which depend on the approximations of the probability of failure in the framework of these methods.

2.6.3.2.1. FORM

The principle of the FORM is simple. It consists of replacing the limit state by a hyper-plan tangent to the design point MPP (see Figure 2.31). The FORM probability of failure is then approximated by the relation:

$$P_f \approx \Phi(-\beta) \qquad [2.85]$$

where $\Phi(\cdot)$ is the centered (see Appendix 4), reduced normal distribution function.

$$\Phi(Z) = \frac{1}{\sqrt{2\pi}} \int_{-\infty}^{Z} e^{-\frac{z^2}{2}} dz, \qquad [2.86]$$

The precision of this method depends upon the nonlinearity of the limit state function.

ILLUSTRATIVE EXAMPLE 2.12.–

Consider an experimental study performed with 10 shafts subjected to a static force F (in compression). Failure occurred at different values of F (Table 2.9). Determine the probability distribution and calculate the probability of failure using the Hasofer–Lind reliability index for a force F = 35KN.

F[KN]	40	45	50	55	60
Number	1	2	4	2	1

Table 2.9. *A set of 10 samples*

The probability distribution of the samples follows the normal distribution (Figure 2.34).

The probability of failure can be determined by calculating the Hasofer–Lind reliability index. To calculate this reliability index, we calculate the mean value and the standard deviation. The mean value is then:

$$m_F = \frac{40\times1+45\times2+50\times4+55\times2+60\times1}{10} = 50 \qquad [2.87]$$

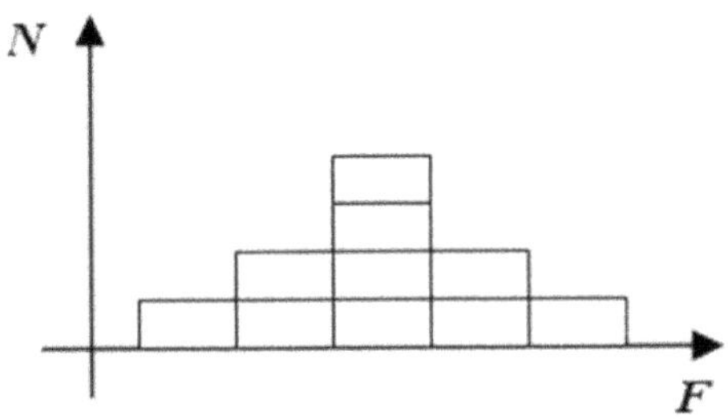

Figure 2.34. *Distribution of samples*

The standard deviation is considered proportional to the mean value ($\sigma_F = 0.1\, m_F = 5$). The normalized variable follows the normal distribution and can be calculated by:

$$u_F = \frac{F - m_F}{\sigma_F} = \frac{35-50}{5} = -3 \qquad [2.88]$$

Since we have a single variable, the reliability index is $\beta = 3$ (see equation [2.79]). Hence, the probability of failure is $P_f = \Phi(-\beta) = 0.13\%$ (see equation [2.87] or Appendix 4).

2.6.3.2.2. SORM

The SORM gives a better estimate of the probability of failure. It consists of replacing the failure surface with a tangential, quadratic hyper-surface with the same curvature as the real surface at the design point MPP (see Figure 2.31) to obtain a better approximation. The SORM probability of failure is hence approximated by the relation:

$$P_f \approx \Phi(-\beta)\prod_{i=1}^{n-1}(1+\beta\kappa_i)^{-1/2} \qquad [2.89]$$

where κ_i is the principal curvature of the limit state at the minimal distance and β is the reliability index obtained by the FORM.

2.6.3.3. *Response surface methods*

The response surface methods (RSMs) are techniques used when the limit state function is implicit, especially when its evaluation requires considerable numerical calculation. These methods aim to construct a mathematical model (called a meta-model) which replaces the limit state function. This function can be defined by regression from numerical experiments. The meta-model thus established replaces the limit state function. Hence, the reliability algorithms or Monte-Carle simulations can be directly applied over the response surface in order to estimate the probability of failure. The response surface methods have two principal advantages:

– the computing time reduction for a limited number of random variables;

– the coupling possibility between the reliability analysis and the optimization algorithms in order to provide an efficient process.

The most common disadvantage lies in the large number of finite element simulations required for a large number of variables.

2.6.4. *Optimality criteria*

Optimization problem [2.78] is equivalent to unconstrained minimization using Lagrange formulation [1.117]:

$$L(\mathbf{u},\lambda_H) = d(\mathbf{u}) + \lambda_H H(\mathbf{u}) \qquad [2.90]$$

with λ_H being the Lagrange multiplier. The derivation of formulation [2.90] leads to:

$$\frac{\partial L}{\partial u_j} = 2u_j + \lambda_H \frac{\partial H}{\partial u_j} = 0 \qquad [2.91]$$

$$\frac{\partial L}{\partial \lambda_H} = H(\mathbf{u}) = 0 \qquad [2.92]$$

This method involves the evaluation of derivatives of the Lagrange formulation in the normalized space. In particular, we do not explicitly know the limit state function *H(***u***)*, and its evaluation is the result of a finite element analysis. However, this analysis demands an enormous amount of computing time, especially for nonlinear and transient problems. Furthermore, the calculation of the normalized gradient $\partial H/\partial u_j$ is not easy because the mechanical analysis is carried out in physical space, and not in the normalized standard space. Calculating the normalized gradient is carried out by applying a chain of rules over the physical gradient $\partial G/\partial y_k$:

$$\frac{\partial H}{\partial u_j} = \frac{\partial G}{\partial y_k} \cdot \frac{\partial T^{-1}(\mathbf{x}, \mathbf{u})}{\partial u_j} \qquad [2.93]$$

These derivatives are generally obtained with finite difference techniques, which demand a high computing time. Post-analysis of a deterministic optimization procedure for reliability analysis reveals two clearly distinct cases.

First case: high reliability level: when the choice of safety factor values for certain parameters is raised, the design cost will be markedly increased because the reliability level may become much greater than the level required for the structure. In this case, we may have a very reliable structure, but with a very high cost.

Second case: low reliability level: a choice of small safety factor values or a bad distribution of these factors entails an unacceptably low reliability level for the structure. For example, Grandhi and Wang

[GRA 98] observed that the reliability index resulting from an optimal deterministic design of a gas turbine blade is β = 0.0053 subject to some uncertainties. This result indicated that the optimum deterministic reliability is very low and must be improved with a probabilistic design.

2.7. Conclusion

Reliability analysis allows us to respect safety or availability criteria imposed during design. The probabilistic approach can be described by 10 steps:

– definition of the failure event;

– choice of mechanical model;

– choice of resistance limit criterion;

– formulation of the mechanical model;

– formulation of the limit state function;

– evaluation of the parameters variability;

– choice of calculation technique;

– calculation of the failure probability;

– sensitivity study of the parameters;

– critical analysis of the results.

This chapter has given a detailed introduction to the reliability concept. Statistical functions and probability distributions have been examined in detail with illustrative examples. Finally, a study of the reliability analysis of structures has been established, ready for its application to biomechanics.

3

Integration of Reliability Concept into Biomechanics

3.1. Introduction

"Biomechanics", by definition, applies the laws of classical mechanics to living organisms. However, if the fundamental notions of forces, moments or even the center of gravity, etc. are still applicable, the geometries, the kinematics, etc. are more complex. Albert Kapandji writes "that in Biomechanics, everything is a curve...". He thereby expresses that no surface is simple, that the axes are no longer represented physically but are virtual, instantaneous and evolutionary [KAP 11]. To this remark, we may add the need to take into account the strong interindividual variations which imply that what is observed in one individual will not necessarily be identical in another individual. A second fundamental difference is the capacity that living beings have to adapt or even "repair" themselves or compensate. These specificities reveal the inherent difficulty of designing a mechanical system (medical device) able to carry out the functions of a biological system.

The risk analysis studies of aging structures and their monitoring currently look to solve many of the challenges, in terms of safety, durability and environmental impact. Essentially based on feedback, most of them prove to be inappropriate. In the structural reliability approach, the influential parameters are considered as random variables. The risk is assessed in the form of a probability and no longer in the form of a binary judgment. This chapter gives some basic elements of reliability of structures. After identifying some causes of the structure failure and the related preventive measures to avoid the failure, we will review the main characteristics of the passive components. We deliberately place ourselves in a user's viewpoint, seeking to minimize the risk of a structure failure and optimize its use. We briefly specify the basis of calculation of the reliability of the structures in order to present biomechanical applications.

Biomechanics is a bioengineering discipline that seeks to apply engineering principles and design procedures to medicine. Biomechanics considers the skeleton and the muscles of the human body as a mechanical structure capable of undergoing a few displacements. It involves human gait analysis and the stress study that tissues and bones are subjected to during accidents. Biomechanical engineering also takes an interest in blood circulation, breathing mechanisms and energy exchanges in the organism. Its field of application ranges from the development of car safety belts to the design and manufacture of heart-lung machines. One of its first developments was the steel lung (a device enabling artificial respiration). Biomechanical engineering also enabled the establishment of the basis for studies on implants and artificial limbs. Prosthetic arms, for instance, operated by a small electric engine and controlled by bioelectric muscular signals have given children born without arms an almost normal appearance and sufficient mobility to ensure their professional and social integration.

Biomechanical research uses geometrical and mechanical modeling and experimental analysis-based approaches close to structural and material mechanics. However, there are a number of barriers both upstream and downstream to characterize and customize the geometry and the materials, which behave differently depending on the remodeling laws (growth, aging, etc.) which still remain less well known, as well as mechanical loading "in service". These key points also concern the validation of both mechanical and physiological models which require the development of quantitative methods for exploration of the living going beyond the traditional field of mechanics and anatomy, while taking into consideration the strong clinical and ethical constraints. In this chapter, the integration of reliability concepts seeks to consider uncertainty on several aspects including loading and properties of bone materials. The description of the behavior of biological tissues and particularly of bone is very complex and remains controversial. Moreover, the change of geometrical parameters precisely in designing prosthesis is studied.

3.2. Origin and categories of uncertainties

Reliability is based on the modeling by some mathematical models of probability and physical aging of the behavior of each component. In order to correctly model the problem, the sources of uncertainties need to be integrated. Three major sources of uncertainties can be identified:

– environmental elements (current and operating expenses) which are at the origin of the various loads applied on a structure;

– the geometrical and mechanical data characterizing the structure;

– the imperfection of the theoretical models used to represent the phenomena (idealizations, imperfections, wrong understanding of the problems, etc.).

There is no way to make a perfectly safe design. Ignoring uncertainty and using safety factors usually lead to designs with inconsistent reliability levels. These uncertainties can be classified into three main categories:

– irreducible uncertainty: irreducible (or physical) uncertainty is due to the inherent randomness in physical phenomena and processes (variability of the height of a wave or the wind speed). It arises during the description of a physical process and still exists even if unlimited data are available. This uncertainty cannot be eliminated or even reduced;

– reducible uncertainty: reducible (or model) uncertainty may happen due to the use of imperfect models to predict outcomes of an action. It results from the simplification of modeling a true physical process and can be minimized by using more sophisticated model;

– statistical uncertainty: statistical (or knowledge) uncertainty is due to the lack of data for modeling uncertainty. Or, it is related to the fitting of a parametric distribution and this uncertainty can be decreased by increasing the number of fitting data points (for example, estimation of a variable from a database containing only a small number of observations).

All these forms of uncertainty are nevertheless processed in the same manner in the analysis irrespective of their origin. Another source of uncertainties is related to human and organizational errors; but it falls outside the scope of this work. The certainty of non-failure does not exist. There are events or a combination of events (overload and bad resistance) which may lead to failure with more or less probability of occurrence. It is only when this probability is considered negligible that a purely deterministic approach can be justified. Several strategies can be used for uncertainty measurements such as: safety factor, worst-case scenario-convex models, Taguchi methods, Fuzzy set methods, probabilistic methods, etc. These strategies lead to a high computing time to compute the probability of failure. An efficient optimization method based on reliability index can be easily implemented and perform the reliability analysis with a reasonable computing time.

Figure 3.1 shows a simplified diagram for design under uncertainty. The design optimization process controls the input parameters (quantified uncertainties) presented by statistical diagram in order to satisfy the required output parameters (calculated uncertainties). The test process is a comparative process between the calculated output and the quantified input until convergence.

3.3. Uncertainties in biomechanics

Biomechanical data are influenced not only by variations which are intrinsic to the subjects [STE 00, YAV 08], but also by the operator performing the experiment [LEA 05, DEL 05] or by the instruments used [CHI 05]. These sources lead to an intrasubject error (for example, average ± standard deviation values for an object) from tests repeated on the same subject and to an intersubject error (for example, a range of values measured for N subjects of interest) coming from different origins or different laboratories [GOR 09].

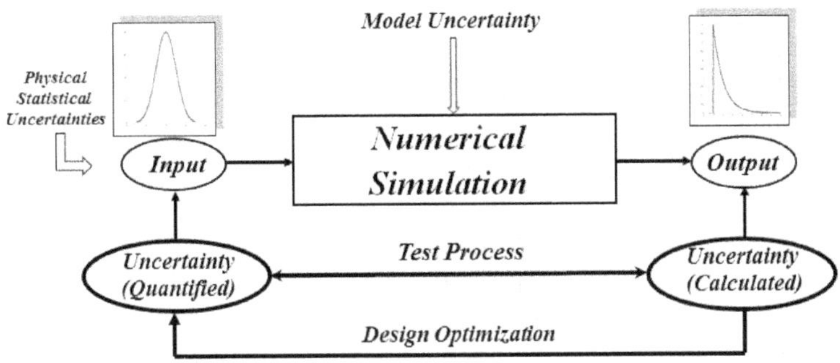

Figure 3.1. *Design under uncertainty*

Moreover, the risk analysis studies on aging structures and their monitoring are currently designed to meet a lot of challenges in terms of safety as well as durability and environmental impact. In this chapter, we present the uncertainties related to loading, geometry and materials.

3.3.1. *Uncertainty in loading*

The various muscle forces are never deterministic in different biomechanical applications. For example, the hip is less mobile than the shoulder. The bending of the hip brings the frontal side of the thigh toward the torso and the extent thereof depends on the knee position. When the thigh touches the trunk, some of the mobility comes from the lumbar spine. The muscles which enable bending are the frontal (anterior) muscles of the thigh and the muscles from inside (interior muscles) the abdomen. The extension is more limited, rendered possible by the back (posterior) muscles of the buttock and the thigh. The abduction, which brings the lower limb outward, is primordial for walking and is mainly taken care of by the buttock muscles which thus take care of the equilibrium of the pelvis. Moreover, during surgery, muscles may be cut or damaged. In such a case, they can never perform to their capacity. This is why the strategy of reliability must be introduced [KHA 16b].

3.3.2. *Uncertainty in geometry*

The design and used materials have greatly evolved since the 1950s especially because of a number of biomechanical studies of the dynamics of joints, in particular during ambulation. Cadaveric studies have brought useful information for prothesis design, but they do not suffice to explain the problems encountered. For example, there are several prothesis sizes, and in order to choose the appropriate size for a patient, it is necessary to perform a real size x-ray image on him/her and a preoperative trace in order to locate the center of the head. Moreover, the prosthesis position can also play a very important role. In general, the integration of uncertainty into the prosthesis geometry and its position are indispensable during the design process [KHA 15a].

3.3.3. *Uncertainty in materials*

The dispersion in the material properties is represented by variations in the materials' characteristics. Uncertainty of mechanical properties can be found in muscular systems and bone composition. The muscular system is responsible for the generation of the forces that enable movement of the bones and joints. Thus, we can get a better understanding of movement through the comprehension of the mechanical properties of the muscular system. The mechanical properties of the muscles are difficult to assess as these structures are made up of complex active or passive tissues [GEN 10]. Understanding the mechanical properties of muscles is crucial in clinical diagnosis. The research on the injuries of the musculoskeletal system and the disorders related to movements, and the application of this knowledge for patient care are at the very core of functional rehabilitation [LEE 07]. Moreover, the description of the behavior of biological tissues and particularly the bone is very complex and remains controversial. For example, when considering the bone as a material with isotropic behavior, we integrate uncertainty on the following four elements:

– Young's modulus;

– Poisson's ratio;

– density;

– yield stress.

In the following sections, we will explain in detail the different aspects and developments of the integration of uncertainty on the bone.

3.4. Bone-related uncertainty

Bones are hard and/or resistant elements and constitute the frame of the human body by supporting the soft parts. These bones (206 of them) fulfill various functions in the human body. On the static plane, they give the body its external shape, support and protect the soft parts, and contain bone marrow. Bone tissue, like cartilage, is a "skeletal tissue", a specialized conjunctive tissue, characterized by the solid nature of the extracellular matrix (ECM). The bone matrix has the particularity of calcification, which makes it opaque to x-rays and allows the study of bones through radiography. The skeleton has three functions:

– mechanical function: bone tissue is one of the most resistant tissues of the organism, capable of bearing mechanical stresses. This gives the bone its role to support the body and to protect the organs;

– metabolic function: bone tissue is a dynamic tissue, constantly restructured under the effect of mechanical pressures, causing the release or storage of mineral salts, and thereby ensuring to a large extent (jointly with the intestine and the kidneys) the control of phosphocalcic metabolism;

– hematopoietic function: the bones contain bone marrow within their medullary spaces, whose stem cells, at the origin of the three blood cell lines, are located near the bone cells. The stromal cells of the bone marrow provide structural and functional support to the hematopoietic cells. Some of them are multipotent stem cells likely to belong to a number of different categories (fibroblasts, chondrocytes, osteoblasts and adipocytes). Besides that, bones detoxify the body by eliminating heavy metals, such as lead and arsenic, as well as other toxins carried by the general circulation.

The bone tissue is made of water (around 1/4 of the bone weight), organic matter (around 1/3 of the bone weight, a major part of which is accounted for by a protein, ossein) and inorganic salts (calcium, phosphorus and magnesium predominate, though we also find iron, sodium, potassium, chlorine and fluorine in small quantities). Two classification methods are used to differentiate the bones of the body. The first classification system is based on the anatomical location of the bone (axial or appendicular), the second on its shape (long, short, flat or uneven). The mandible is part of the 80 axial bones which are distributed along the central, vertical axis of the body.

3.4.1. *Bone behavior law*

Robert Hooke (1635–1703) was an English physicist. He was one of the greatest experimental scientists of the 17th Century and the inventor of the first telephone in 1665.

Augustin-Louis Cauchy (1789–1857) was a French mathematician. He was interested in problems related to complex analysis, algebra and probabilities.

Thomas Young (1773–1829) was an English doctor and physicist. He practiced medicine throughout his life, but he is known above all for his definition of the Young's modulus in material science and for his double-slit experiment in optics, in which he highlighted and interpreted the light interference phenomenon.

Siméon Denis Poisson (1781–1840) was a French mathematician, geometrician and physicist. He introduced the ratio which bears his name. This ratio is one of the fundamental elements of the theory of elasticity. He concerned himself with the definite integral, Fourier transform and the theory of probabilities (Poisson distribution).

Hooke's law was generalized by Cauchy, who proposed to express each component of the stress tensor as a linear function of the components of the strain tensor. Hooke's law is therefore often written using the following formula:

$$\boldsymbol{\sigma} = \mathbf{C}.\boldsymbol{\varepsilon} \qquad [3.1]$$

where **C** is a fourth-order tensor called rigidity tensor or elasticity tensor (the covariant components of this tensor are C_{ijkl}). The rigidity tensor contains the set of elastic characteristics of the material. Similarly, the deformations are linearly related to the stresses by an inverse relationship:

$$\boldsymbol{\varepsilon} = \mathbf{S}.\boldsymbol{\sigma} \qquad [3.2]$$

where **S** is the compliance tensor or elastic compliance tensor of the material (its covariant components are S_{ijkl}). The **C** and **S** tensors have 81 components in principle (each index varies from 1 to 3). However, we have seen that Cauchy's stress and strain tensors are symmetrical. Therefore each of them has only six independent components, and their linear relation can then be established with the help of just 36 terms. The following formula is often used, in an orthonormal pinpoint, to relate the stress and strain components:

$$\begin{Bmatrix} \sigma_{11} \\ \sigma_{22} \\ \sigma_{33} \\ \sigma_{23} \\ \sigma_{13} \\ \sigma_{12} \end{Bmatrix} = \begin{bmatrix} C_{1111} & C_{1122} & C_{1133} & C_{1123} & C_{1131} & C_{1112} \\ C_{2211} & C_{2222} & C_{2233} & C_{2223} & C_{2231} & C_{2212} \\ C_{3311} & C_{3322} & C_{3333} & C_{3323} & C_{3331} & C_{3312} \\ C_{2311} & C_{2322} & C_{2333} & C_{2323} & C_{2331} & C_{2312} \\ C_{3111} & C_{3122} & C_{3133} & C_{3123} & C_{3131} & C_{3112} \\ C_{1211} & C_{1222} & C_{1233} & C_{1223} & C_{1231} & C_{1212} \end{bmatrix} \begin{Bmatrix} \varepsilon_{11} \\ \varepsilon_{22} \\ \varepsilon_{33} \\ 2\varepsilon_{23} \\ 2\varepsilon_{13} \\ 2\varepsilon_{12} \end{Bmatrix} \qquad [3.3]$$

with the condition $C_{ijkl} = C_{ijlk} = C_{jikl} = C_{jilk}$. The matrix components presented in the previous relationship are often denoted as C_{IJ}, with I and J varying from 1 to 6. We can then write:

$$\begin{Bmatrix} \varepsilon_{11} \\ \varepsilon_{22} \\ \varepsilon_{33} \\ \varepsilon_{23} \\ \varepsilon_{13} \\ \varepsilon_{12} \end{Bmatrix} = \begin{bmatrix} C_{11} & C_{12} & C_{13} & C_{14} & C_{15} & C_{16} \\ C_{12} & C_{22} & C_{21} & C_{24} & C_{25} & C_{26} \\ C_{13} & C_{23} & C_{33} & C_{34} & C_{35} & C_{36} \\ C_{14} & C_{24} & C_{34} & C_{44} & C_{45} & C_{46} \\ C_{15} & C_{25} & C_{35} & C_{45} & C_{55} & C_{56} \\ C_{16} & C_{26} & C_{36} & C_{46} & C_{56} & C_{66} \end{bmatrix} \begin{Bmatrix} \sigma_{11} \\ \sigma_{22} \\ \sigma_{33} \\ \sigma_{23} \\ \sigma_{13} \\ \sigma_{12} \end{Bmatrix} \quad [3.4]$$

For orthotropic behavior (i.e. mandible, etc.), we consider nine constants:

$$\begin{aligned} & E_1, E_2, E_3, \\ & G_{12}, G_{23}, G_{31} \\ & \nu_{12} = \nu_{21}, \ \nu_{23} = \nu_{32}, \ \nu_{13} = \nu_{31}, \end{aligned} \quad [3.5]$$

The material form is then presented by the following formula:

$$\begin{Bmatrix} \varepsilon_{11} \\ \varepsilon_{22} \\ \varepsilon_{33} \\ \varepsilon_{23} \\ \varepsilon_{13} \\ \varepsilon_{12} \end{Bmatrix} = \begin{bmatrix} \frac{1}{E_1} & -\frac{\nu_{21}}{E_1} & -\frac{\nu_{31}}{E_1} & 0 & 0 & 0 \\ -\frac{\nu_{12}}{E_2} & \frac{1}{E_2} & -\frac{\nu_{32}}{E_2} & 0 & 0 & 0 \\ -\frac{\nu_{13}}{E_3} & -\frac{\nu_{23}}{E_3} & \frac{1}{E_3} & 0 & 0 & 0 \\ 0 & 0 & 0 & \frac{1}{G_{23}} & 0 & 0 \\ 0 & 0 & 0 & 0 & \frac{1}{G_{31}} & 0 \\ 0 & 0 & 0 & 0 & 0 & \frac{1}{G_{12}} \end{bmatrix} \begin{Bmatrix} \sigma_{11} \\ \sigma_{22} \\ \sigma_{33} \\ \sigma_{23} \\ \sigma_{13} \\ \sigma_{12} \end{Bmatrix} \quad [3.6]$$

For transversal isotropic behavior (i.e. femur, tibia, etc.), the number of constants is decreased so that deformation can be in two dimensions (Figure 3.2).

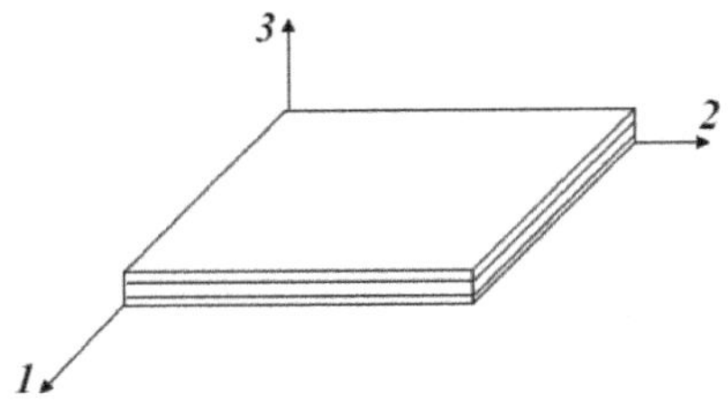

Figure 3.2. *Description of transversal isotropic behavior by a shell*

For transversal isotropic behavior, we consider six constants as follows:

$$\begin{array}{l} E_1 = E_2,\ E_3 \\ \nu_{31} = \nu_{32},\ \nu_{12} \\ G_{31} = G_{32},\ G_{12} = \dfrac{E_1}{2(1+\nu_{12})} \end{array} \quad [3.7]$$

The material form is then presented by the formula:

$$\begin{Bmatrix} \varepsilon_{11} \\ \varepsilon_{22} \\ \varepsilon_{33} \\ \varepsilon_{23} \\ \varepsilon_{13} \\ \varepsilon_{12} \end{Bmatrix} = \begin{bmatrix} \dfrac{1}{E_1} & -\dfrac{\nu_{12}}{E_1} & -\dfrac{\nu_{31}}{E_1} & 0 & 0 & 0 \\ -\dfrac{\nu_{12}}{E_1} & \dfrac{1}{E_1} & -\dfrac{\nu_{31}}{E_1} & 0 & 0 & 0 \\ -\dfrac{\nu_{31}}{E_3} & -\dfrac{\nu_{31}}{E_3} & \dfrac{1}{E_3} & 0 & 0 & 0 \\ 0 & 0 & 0 & \dfrac{1}{G_{31}} & 0 & 0 \\ 0 & 0 & 0 & 0 & \dfrac{1}{G_{31}} & 0 \\ 0 & 0 & 0 & 0 & 0 & \dfrac{2(1+\nu_{12})}{E_1} \end{bmatrix} \begin{Bmatrix} \sigma_{11} \\ \sigma_{22} \\ \sigma_{33} \\ \sigma_{23} \\ \sigma_{13} \\ \sigma_{12} \end{Bmatrix} \quad [3.8]$$

For isotropic behavior, the number of constants is decreased but the deformation is still described in three-dimensions (3D) (Figure 3.3).

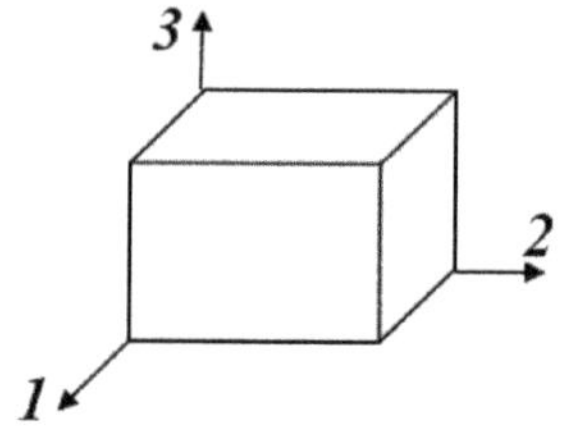

Figure 3.3. *Description of isotropic behavior by a cube*

For isotropic behavior, we consider three constants as follows:

$$\begin{aligned} &E = E_1 = E_2 = E_3 \\ &\nu = \nu_{31} = \nu_{32} = \nu_{12} \\ &G = G_{31} = G_{32} = G_{12} = \frac{E}{2(1+\nu)} \end{aligned} \quad [3.9]$$

The material form is then presented by the following formula:

$$\begin{Bmatrix} \varepsilon_{11} \\ \varepsilon_{22} \\ \varepsilon_{33} \\ \varepsilon_{23} \\ \varepsilon_{13} \\ \varepsilon_{12} \end{Bmatrix} = \begin{bmatrix} \frac{1}{E} & -\frac{\nu}{E} & -\frac{\nu}{E} & 0 & 0 & 0 \\ -\frac{\nu}{E} & \frac{1}{E} & -\frac{\nu}{E} & 0 & 0 & 0 \\ -\frac{\nu}{E} & -\frac{\nu}{E} & \frac{1}{E} & 0 & 0 & 0 \\ 0 & 0 & 0 & \frac{2(1+\nu)}{E} & 0 & 0 \\ 0 & 0 & 0 & 0 & \frac{2(1+\nu)}{E} & 0 \\ 0 & 0 & 0 & 0 & 0 & \frac{2(1+\nu)}{E} \end{bmatrix} \begin{Bmatrix} \sigma_{11} \\ \sigma_{22} \\ \sigma_{33} \\ \sigma_{23} \\ \sigma_{13} \\ \sigma_{12} \end{Bmatrix} \quad [3.10]$$

3.4.2. *Contribution to the characterization of the bone's mechanical properties*

Several experiments *in vitro* have been performed in order to characterize bone material properties at the Solid Mechanics Laboratory at Lund University. This study can be realized considering the two levels already investigated:

– *Macrostructure*: in this case, we consider that the bone material is homogeneous in order to simplify the numerical calculation. The tests are carried out just to validate the previous numerical results of our new model. The evaluation of fracture risks and the integration of different uncertain data are important steps to evaluate the reliability level of structures. Reliability-based methods lead to indicators of the confidence that we can place on a choice of design and sizing with regard to potential failure scenarios, taking into account modeling uncertainties [KHA 15a].

– *Microstructure*: osteoporosis is a major public health problem which can increase fracture risk [SHO 15]. The reduction of the bone mass (osteopenia) with advancing age is an unavoidable physiological phenomenon. We consider porosity during the structural behavior modeling of the compact bone. In this case, we study the diffusion and rigidity of small samples. Mechanical behavior is also identified by the technique (digital image correlation (DIC)).

Figure 3.4 shows three DIC images of the deformation of a cortical bone cube of a cow to reach fissure which is illustrated in Figure 3.4(c).

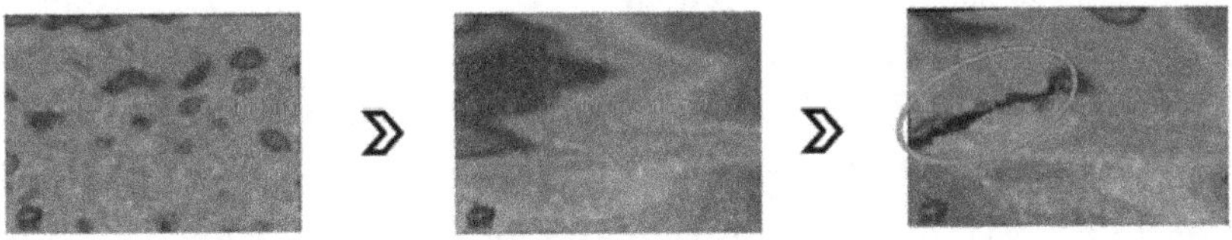

a) b) c)

Figure 3.4. *Three successive images of DIC compression technique*

3.5. Bone developments and formulations

3.5.1. *Current formulation*

Generally, there is a relationship between bone density and its mechanical properties. According to Keller [KEL 94], the modulus of elasticity (Young's modulus) and the yield stress in compression are presented by the following formula:

$$E = 10500\rho_\alpha^{2.57\pm0.04} \qquad [3.11]$$

$$\sigma_C = 117\rho_\alpha^{1.93\pm0.04} \qquad [3.12]$$

where ρ_α is the ash density. For trabecular (or cancellous) and cortical bones, an interval of this density ρ_α = $(0.03 \div 1.22 g/cm^3)$ can be found in the literature [DOB 04]. This leads to an interval of elasticity modulus: $E = (1.47 \div 17643.65\, MPa)$ and an interval of yield stress in compression: $\sigma_C = (0.15 \div 173.12\, MPa)$. However, Keyak *et al.* [KEY 94] have developed other formulations to relate the ash density with the elasticity modulus and yield stress in compression for the trabecular bone considering the following formula:

$$E = 33900\rho_\alpha^{2.2}, \text{ if } \rho_\alpha \leq 0.27\ g/cm^3 \qquad [3.13]$$

$$\sigma_C = 137\rho_\alpha^{1.88}, \text{ if } \rho_\alpha \leq 0.317\ g/cm^3 \qquad [3.14]$$

We can simply test the difference between the two models. We consider that ash density is equal to $\rho_\alpha = 0.27\ g/cm^3$. According to Keller's model [KEL 94], the modulus of elasticity is therefore $E = (281.72 \div 314.76)\, MPa$ and the yield stress in compression is $\sigma_C = (7.62 \div 8.52)\, MPa$. However, according to the model of Keyak *et al.* [KEY 94], the modulus of elasticity is $E = 1605.7\, MPa$ and the yield stress in compression is $\sigma_C = 10.11\, MPa$. For the same density, we can find a big difference between the results. For this reason, we need a formulation with a good level of accuracy. To find this formulation, we can first generalize the formulations and then use optimization technology to reduce the error of the developed formulation.

3.5.2. *Generalized formulation*

The different equations relating mechanical properties with ash density can be written with the following formula:

$$E = A_E \rho_\alpha^{n_E} \tag{3.15}$$

$$\sigma_C = A_\sigma \rho_\alpha^{n_\sigma} \tag{3.16}$$

In this case, we consider that A_E and A_σ are the constants of proportionality and that n_E and n_σ are the proportionality indices. Since the elasticity modulus and the yield stress in compression are related to ash density, we can write the yield stress in compression as a function of the modulus of elasticity. First, equation [3.15] can be written as follows:

$$\frac{E}{A_E} = \rho_\alpha^{n_E} \tag{3.17}$$

Considering the logarithm of this equation, we obtain:

$$\frac{1}{n_E} \ln\left(\frac{E}{A_E}\right) = \ln \rho_\alpha \tag{3.18}$$

We repeat the same thing with equation [3.16]:

$$\frac{\sigma_C}{A_\sigma} = \rho_\alpha^{n_\sigma} \tag{3.19}$$

Considering the logarithm of this equation, we obtain:

$$\frac{1}{n_\sigma}\ln\left(\frac{\sigma_C}{A_\sigma}\right)=\ln\rho_\alpha \qquad [3.20]$$

Eliminating the logarithm of the ash density, we obtain:

$$\frac{1}{n_\sigma}\ln\left(\frac{\sigma_C}{A_\sigma}\right)=\frac{1}{n_E}\ln\left(\frac{E}{A_E}\right) \qquad [3.21]$$

This equation can be written as follows:

$$\sigma_C = A_\sigma e^{\left(R_{\sigma/E}\ln\left(\frac{E}{A_E}\right)\right)} \qquad [3.22]$$

where $R_{\sigma/E} = n_\sigma/n_E$ is the rate of the indices. The different studies are based on the yield stress in compression, but it actually depends on the tension. Thus, the formulation in tension can be written using the following formula:

$$\sigma_T = R_{T/C}.A_\sigma e^{\left(R_{\sigma/E}\ln\left(\frac{E}{A_E}\right)\right)} \qquad [3.23]$$

The tension/compression rate $R_{T/C}$ is 0.5–0.7 for the cortical bone and 0.7–1 for the trabecular bone [DOB 04].

3.5.3. *Optimized formulation*

The constants of equation [3.22] can be determined by considering some experimental results. It is also easy to find the rate of indices $R_{\sigma/E} = n_\sigma/n_E$ by using two experimental points: (σ_{C_i}, E_i) and $(\sigma_{C_{i+1}}, E_{i+1})$. Thus, equation [3.22] can be written as follows:

$$\ln\sigma_{C_i} - \ln A_\sigma = R_{\sigma/E}\left(\ln E_i - \ln A_E\right) \qquad [3.24]$$

$$\ln\sigma_{C_{i+1}} - \ln A_\sigma = R_{\sigma/E}\left(\ln E_{i+1} - \ln A_E\right) \qquad [3.25]$$

This way, the ratio of exponents can be written as follows:

$$R_{\sigma/E} = \frac{\ln \sigma_{C_i} - \ln \sigma_{C_{i+1}}}{\ln E_i - \ln E_{i+1}} \quad [3.26]$$

In order to get all constants with an optimum fitting curve, an iterative (optimization) method can be used for at least three given experiment points: (σ_{C_i}, E_i), $(\sigma_{C_{i+1}}, E_{i+1})$ and $(\sigma_{C_{i+2}}, E_{i+2})$. Let us consider the three experimental results for cortical bone presented in Table 3.1. Three logarithmic equations can be formulated as follows:

$$5.257 - \ln A_\sigma = R_{\sigma/E}\left(9.741 - \ln A_E\right) \quad [3.27]$$

$$5.278 - \ln A_\sigma = R_{\sigma/E}\left(9.764 - \ln A_E\right) \quad [3.28]$$

$$5.323 - \ln A_\sigma = R_{\sigma/E}\left(9.793 - \ln A_E\right) \quad [3.29]$$

The three logarithmic equations are optimized in order to find the constant values with an optimum fitness. The resulting optimum constants are presented in Table 3.1. Here, the optimized formulation in compression can be written as follows:

$$\sigma_C = 129.5\left(\frac{E}{11142}\right)^{0.94} \quad [3.30]$$

Thus, the optimized formulation in tension can be written as follows:

$$\sigma_T = 129.5 R_{T/C}\left(\frac{E}{11142}\right)^{0.94} \quad [3.31]$$

According to Keaveny *et al.* [KEA 01], the ultimate tensile strength equals 132 MPa, while the compressive strength equals 192 MPa for cortical bone. This way, the tensile/compressive ratio is acceptable: $R_{T/C} \approx 0.69 \in \left[0.5 \div 0.7\right]$

Author	Experiments		Optimized constants			Kharmanda's model	Keller's model
	E	σ_C	A_E	A_σ	$R_{\sigma/E}$	σ_C	σ_C
Keaveny *et al.* [KEA 01]	17,000	192	11,142	129.5	0.94	192.64	167.93
Gibbons *et al.* [GIB 76]	17,400	196				196.90	170.89
Reilly *et al.* [REI 75]	17,900	205				202.21	174.56

Table 3.1. *Experimental results for cortical bone and model validation*

When using the new model for both cortical and trabecular layers (Tables 3.1 and 3.2), the results seem to be much closer to the experimental values than those produced by the classical model of Keller [KEL 94].

Author	Experiments		Optimized constants			Kharmanda's model	Keller's model
	E	σ_C	A_E	A_σ	$R_{\sigma/E}$	σ_C	σ_C
Aleixo *et al.* [ALE 13]	761.0	10.53	11,142	129.5	0.94	10.39	16.30
Schoenfeld *et al.* [SCH 74]	965	13.5				12.99	19.48

Table 3.2. *Experimental results for trabecular bone and model validation*

3.5.4. *Extension to orthotropic behavior formulation*

There are a few types of bones which cannot be treated as isotropic materials. For example, the mandible is subjected to loading in several directions and mechanical properties are different in three Cartesian directions. The behavior is thus considered orthotropic. In this case, we need to develop a formulation that will relate Young's modulus with yield stress of the bone [KHA 16c].

Let us consider the classical symmetric compliance matrix of an orthotropic material presented by the following form:

$$\mathbf{S}=\begin{bmatrix} \frac{1}{E_x} & -\frac{\nu_{yx}}{E_y} & -\frac{\nu_{zx}}{E_z} & 0 & 0 & 0 \\ -\frac{\nu_{xy}}{E_x} & \frac{1}{E_y} & -\frac{\nu_{zy}}{E_z} & 0 & 0 & 0 \\ -\frac{\nu_{xz}}{E_x} & -\frac{\nu_{yz}}{E_y} & \frac{1}{E_z} & 0 & 0 & 0 \\ 0 & 0 & 0 & \frac{1}{G_{xy}} & 0 & 0 \\ 0 & 0 & 0 & 0 & \frac{1}{G_{xz}} & 0 \\ 0 & 0 & 0 & 0 & 0 & \frac{1}{G_{yz}} \end{bmatrix} \quad [3.32]$$

To determine the Young's modulus and Poisson's ratio of an equivalent isotropic medium, strain energies involved in both materials by the elementary stress states have to be compared. Strain energy density is calculated using the following expression:

$$W=\frac{1}{2}\sum_{i,j=1}^{3}\varepsilon_{ij}\sigma_{ij} \quad [3.33]$$

Using the generalized Hooke's law:

$$\varepsilon_{ij}=\sum_{i,j=1}^{3}S_{ijkl}\sigma_{kl} \quad [3.34]$$

Therefore, formulation [3.33] can be written as follows:

$$W=\frac{1}{2}\sum_{i,j,k,l=1}^{3}\sigma_{ij}S_{ijkl}\sigma_{kl} \quad [3.35]$$

The equivalence of strain energies is then carried out for two elementary loadings, i.e. hydrostatic pressure and 3D shearing. Corresponding stress states are as follows:

– $\sigma_{11} = \sigma_{22} = \sigma_{33} = \sigma$ for hydrostatic pressure (preserving the material element shape);

– $\sigma_{11} = \sigma_{22} = \sigma_{33} = \sigma$ for 3D shearing (preserving the material element volume).

The corresponding expressions of strain energy densities can then be derived for orthotropic and isotropic materials. For the hydrostatic pressure, we have:

$$W_{Hydrostatic}^{Orthotropic} = \frac{1}{2}\sigma^2\left(\frac{1}{E_x}+\frac{1}{E_y}+\frac{1}{E_z}-2\frac{\nu_{yx}}{E_y}-2\frac{\nu_{zy}}{E_z}-2\frac{\nu_{xz}}{E_x}\right) \quad [3.36]$$

$$W_{Hydrostatic}^{Isotropic} = \frac{1}{2}\sigma^2\frac{3}{E}(1-2\nu) \quad [3.37]$$

For the 3D shearing, we have:

$$W_{Shear}^{Orthotropic} = \frac{1}{2}\tau^2\left(\frac{1}{G_{xy}}+\frac{1}{G_{yz}}+\frac{1}{G_{zx}}\right) \quad [3.38]$$

$$W_{Shear}^{Isotropic} = \frac{1}{2}\tau^2\frac{3}{G} \quad [3.39]$$

The equivalence request concerning the energies for these two loads leads to:

$$\left(\frac{1}{E_x}+\frac{1}{E_y}+\frac{1}{E_z}-2\frac{\nu_{yx}}{E_y}-2\frac{\nu_{zy}}{E_z}-2\frac{\nu_{xz}}{E_x}\right)=\frac{3}{E}(1-2\nu)=\frac{1}{K} \quad [3.40]$$

$$\frac{1}{G_{xy}}+\frac{1}{G_{yz}}+\frac{1}{G_{zx}}=\frac{3}{G} \quad [3.41]$$

or else, we can define the bulk modulus for an equivalent isotropic material by the following formula :

$$K = \frac{E_x E_y E_z}{(1-2\nu_{xz})E_y E_z + (1-2\nu_{yx})E_x E_z + (1-2\nu_{zy})E_x E_y} \qquad [3.42]$$

and the shear modulus using the following formula:

$$G = \frac{3G_{xy}G_{yz}G_{zx}}{G_{yz}G_{zx} + G_{xy}G_{zx} + G_{xy}G_{yz}} \qquad [3.43]$$

Thus, the equivalent Young's modulus is written as follows:

$$E = \frac{9KG}{3K+G} \qquad [3.44]$$

Furthermore, the equivalent Poisson's ratio is written as follows:

$$\nu = \frac{3K-2G}{2(3K+G)} \qquad [3.45]$$

In this case, we can calculate the yield stress in compression and tension using equations [3.22] and [3.23].

3.6. Characterization by experimentation of the bone's mechanical properties

In order to determine the yield stress in compression, we can use compression tests, however, the yield stress in tension can be determined by using the bending test because the bone is too fragile to be processed by tensile tests. We begin with the bending test at three points in order to validate the optimized formulation ([3.30] and [3.31]). We subsequently determine the relationship between the tensile strength and the displacement rate (speed) by compression tests.

3.6.1. *Characterization by bending test*

The validation of equation [3.31] for the yield stress in tension was performed by bending tests at the Solid Mechanics Laboratory at Lund University (Figure 3.5).

Figure 3.5. *Bending test at three points*

Figure 3.6 shows a sample of a cow's cortical bone with dimensions: $47.5 \times 15.2 \times 7.9\ mm$. The rate of displacement (speed) is equal to: $R = 0.5\ mm/\text{min}$ (Figure 3.6). The density is $\rho = 1.66$ g/cm^3 and the breaking force is $P = 1351\ N$. By applying the calculation laws of the bending test at three points, we find that the modulus of elasticity (Young) is $E = 11.9\ GPa$ and the yield stress is $\sigma_T = 101.5\ MPa$, respectively. The optimized formulation [3.31] leads to a yield stress of $\sigma_T = 96.4\ MPa$. This value is not far off the experimental value by taking into account the various errors generated by experimental measurements and the test protocol [KHA 15b].

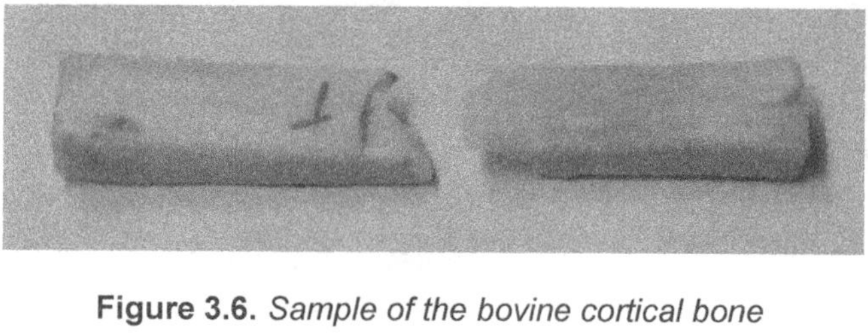

Figure 3.6. *Sample of the bovine cortical bone*

3.6.2. *Characterization by compression test*

In this test, we model the relationship between the tensile strength and loading speed. The interval of displacement rate is $R = [0.5 - 5]$ mm/min. The tests are performed for several samples. Figure 3.7 also shows the compression experimentation at the Solid Mechanics Laboratory at Lund University.

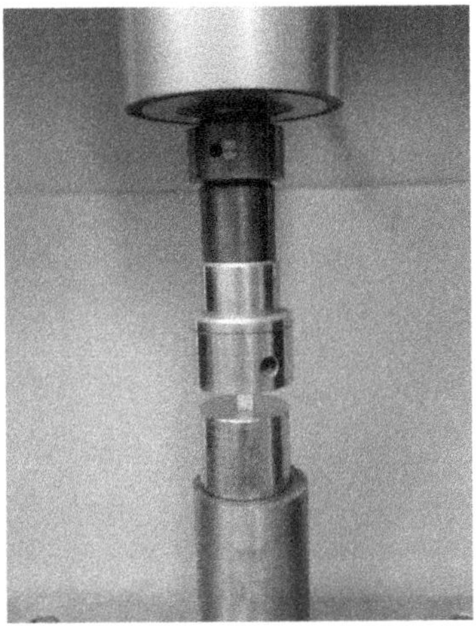

Figure 3.7. *Compression test*

Figure 3.8 shows the relationship between tensile strength and the displacement rates for samples with the same density ($\rho = 1.7 g/cm^3$) and in the axial direction of the femur bone. We can conclude that tensile strength increases during an increase in the displacement rate in the axial direction. Otherwise, long bones (femur, tibia, etc.) can offer resistance against dynamic loading. In general, the loading on these types of bones is axial. For this reason, we can treat it like an isotropic material but in the other types of bones like the mandible, the behavior of anisotropic materials is indispensable. In this book, we take the uncertainties into account in order to find a solution to overcome this difficulty or the lack of information.

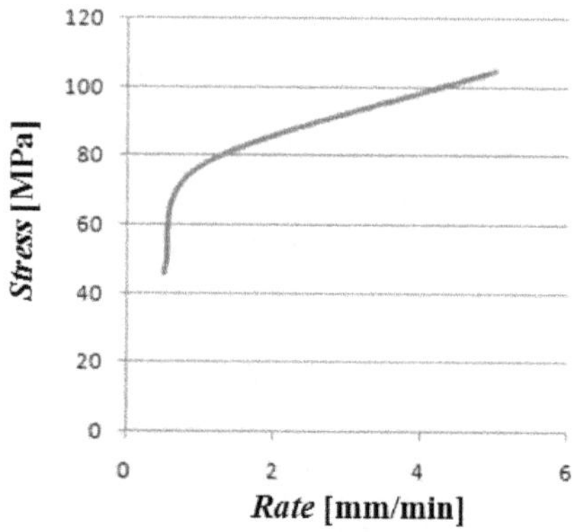

Figure 3.8. *Relation stress/rate*

3.7. Conclusion

The integration of reliability analysis has become indispensable in a few disciplines: aeronautics, nuclear power, etc. In this chapter, we present the integration of reliability concept into biomechanics. The various uncertainties can come from lack of information or variation of the model. The uncertainties coming from a lack of information about the parameters of models and their variations can be found in loading, geometry and/or materials properties. The change in the properties of materials in the bone poses a problem during the analysis of reliability. Finally, we have presented the developments of formulations characterizing the bone's mechanical properties based on experimentation trials.

4

Reliability Analysis of Orthopedic Prostheses

4.1. Introduction to orthopedic prostheses

Orthopedic prostheses are used to replace a deficient or painful joint. They refer to the prostheses of the knee, hip, shoulders, elbow, etc. In industrialized countries, the progress made by science has increased lifespans considerably. Today, we are faced with a serious problem concerning old age: arthropathies and especially arthrosis. The hip joint is one of the joints most affected by arthrosis. In fact, it is a highly mobile joint used essentially for walking, but it is also a considerable point of conflict between gravity and reactive thrust of the ground. An American report of the National Institute for Health [WEI 03], found that more than 120,000 artificial hip prostheses are implanted annually in the United States. In France, this number exceeds 50,000 prostheses per year. Though new technologies and surgical techniques have considerably reduced the risks associated with prosthetic hip replacement and have improved the immediate and long-term outcomes, some questions still remain. In fact, we are still pondering over prosthetic design as well as on the biological response of the bone/implant environment. It is therefore difficult for orthopedic surgeons to make a preoperative plan for a total hip prosthesis (THP). Over the last few years, numerical simulation has become an essential tool in biomechanics for understanding the behavior of human beings subjected to a traumatism. Though this scope of application constitutes a number of studies in the biomedical domain, it also holds great interest insofar as its use for developing orthopedic prostheses is concerned. These applications require the creation of finite element models of the human being and different prosthetic profiles in order to test them in various configurations of mechanical solicitations. In order to optimize the geometrical acquisition and the meshing model of each element, the use of software which makes it possible to automate these two steps has become an important issue.

Therefore, numerical modeling, which is less expensive, is intended to help in the diagnosis of orthopedic surgeons with respect to the implant choice, by making it possible to foresee the long-term behavior of the prosthesis, even after 20 years, a perspective that surgeons do not currently possess.

Reliability analysis can be done by taking into consideration the hazards related to loading, materials and/or geometry. In the context of loading, the damage mechanisms can contribute to the design and improvement of prostheses. Moreover, the bone's mechanical properties can be changed after the surgery. First, the bone is a living "material" whose characteristics evolve depending on the weather, bone type, person's gender, his/her age, his/her health, etc. Furthermore, the modification of mechanical actions, caused for example by the fixation of an implant, can lead to a radical change in its quality. Bone remodeling and adaptation theories attempt to describe and foresee these changes. A simple and pedagogical application of reliability analysis is made by considering the uncertainty on the geometry. Next, a detailed application on the hip prosthesis is carried out by taking into consideration the uncertainty on the materials in order to assess the reliability level of the prosthesis.

4.1.1. *History of prostheses*

Artificial limbs have been used right from Antiquity. In 1885, a specimen was discovered in a tomb in Capua, Italy, along with other remains dating back to 300 BC. The famous artificial hand built in 1509 for the German knight Gőtz von Berlichingen, also known as Götz of the Iron Hand, weighed around 1.4 kg and had articulated fingers which were arranged in such a manner that the prosthetic hand was capable of holding objects such as a saber or a lance. This hand, which is still functional, is exhibited in the Nuremberg museum. In the early 19th Century, a German prosthetist crafted an artificial hand equipped with fingers which could bend and stretch all by itself and could even close in order to be able to hold light objects such as a pencil, a handkerchief or a hat. In 1851, a French prosthetist invented an artificial arm fixed with a wooden hand, attached to a leather sleeve which fixes the stump snugly. The fingers were half shut, the thumb pivoted on a nail and could rest firmly against the other fingers due to a thick and invisible rubber strip; a mechanism fixed on the opposite shoulder allowed us to activate the thumb. The same inventor also developed a wooden leg which attempted to reproduce a natural gait and was capable of extending its pace.

4.1.2. *Evolution of prostheses*

Prostheses production veritably became a science subsequent to the mutilations caused by the different wars which left their imprint on the 20th Century. Artificial legs articulated at the knee and ankle, as well as hip prostheses, were able to return a practically normal mobility and gait to the mutilated soldiers. Over the years, technological advances and the use of metal alloys have made it possible to produce artificial limbs that are at once light and resistant. Synthetic polymers for example, which are now used, give a natural appearance to the material which covers certain prostheses.

4.1.3. *Examples of orthopedic prostheses*

An orthopedic prosthesis is a mechanical device adapted to reproduce the shape and as much as possible the function of a lost or missing limb. Orthopedic prostheses are used to replace a deficient or painful joint. The replacement of the limb with an artificial substitute is done by a specialized surgeon. They refer to the prostheses of the knee, hip, shoulders, elbow, etc.

The results of an arthroplasty are based, in equal parts, on the choice of the prosthesis and the mastery of the surgical technique for the associated implantation. Economically, the direct cost of the prosthesis seems secondary and should be related with the costs of the operative and postoperative periods on the one hand, and the services rendered to the patient on the other hand.

4.2. Reliability analysis of the intervertebral disk

First, the basics of the anatomy of the lumbar spine are presented, then the intervertebral disk at the lumbar level will be used to apply the reliability analysis.

4.2.1. *Functional anatomy*

The projection of the spine in the frontal plane is straight in a healthy individual (except in the case of some pathologies) and curved in the sagittal plane (Figure 4.1). Four curvatures can be distinguished:

– cervical curvature also called cervical lordosis;

– dorsal curvature also named dorsal or thoracic kyphosis;

– lumbar curvature called lumbar lordosis;

– sacral curvature which is fixed because of the soldering of the sacral vertebrae.

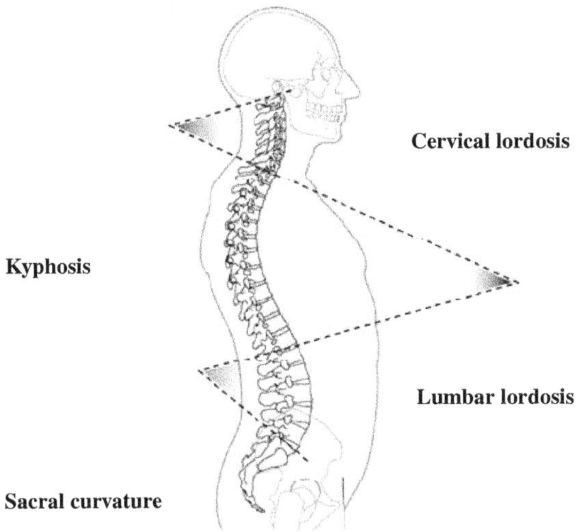

Figure 4.1. *Sagittal curvatures of the spine*

In this chapter, we will focus on the lumbar in order to perform reliability studies.

4.2.2. *The lumbar functional spinal unit*

4.2.2.1. *Description*

The lumbar functional spinal unit (FSU), also called "lumbar motion segment", is defined as a set of two adjacent vertebrae and their relating elements (deformable elements). It can be broken down into two segments: the segment known as passive and the segment known as mobile (deformable) according to Kapandji and Palombi [KAP 07, PAL 07]. The passive segment is made up of both the vertebrae (upper and lower). The mobile segment is made up of:

– anterior part: intervertebral disk (IVD), anterior longitudinal ligament (ALL) and posterior longitudinal ligament (PLL);

– posterior part (posterior arch): intervertebral foramen, zygapophyseal joints and ligaments [ligamentum Flavum (LF), intertransverse, interspinous and supraspinous].

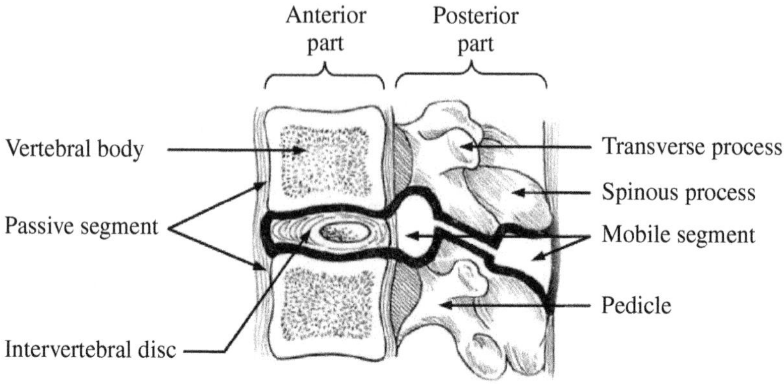

Figure 4.2. *Disk prosthesis in situ*

The mobile segment allows the movements of the functional unit and the entire lumbar. The PLL and LF are taut in flexion and the ALL in extension. They have a viscoelastic behavior, which means that their stiffness increases with the loading speed. These longitudinal ligaments maintain a compressive force along the spine, regardless of its position in the sagittal plane. They thus minimize the tensile forces which could damage the superficial fibers of the IVD.

4.2.2.2. *The intervertebral disk*

The IVD is an avascular structure composed of 65–90% water [RAN 04]. It contains two parts: in its center, the nucleus pulposus (NP) and in the periphery, the annulus fibrosus (AF) (Figure 4.2). The NP and the inner part of the AF are in direct contact with the cartilage (hyalin) of the vertebral endplates also called vertebral cartilaginous endplates (VCP). The NP, which occupies around 50% of the IVD volume, is located slightly to the rear of the center of the disk (in the sagittal plane) and moves during movements of the FU. It is inextensible, incompressible, deformable and highly hydrophilic. The NP is rich in proteoglycans, macromolecules which get attached to water. The AF, with a firm and elastic lamellar fibrous structure, maintains the NP in position in the intervertebral space. It is richer in collagen (types 1, 2 and 4 mainly) but less rich in water content than the NP.

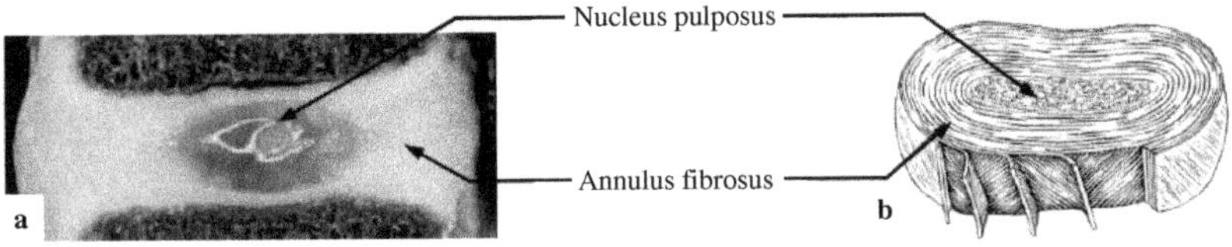

Figure 4.3. *Intervertebral disk: a) Sagittal section of a healthy disk; b) illustration of the disk*

The IVD's architecture changes with age. It is the seat of gradual maturation till adulthood (around 30 years) followed by aging with variable intensity depending on the individual [COT 00]. While resisting complex external mechanical stresses, its function is to transfer load and ensure multi-directional mobility of the FSU. The thickness of the lumbar IVDs varies from 10 to 15 mm and increases from L1-L2 to L4-L5. A high disk index (ratio between the heights of the IVD and the vertebral body equal to 1/5) favors mobility [RAN 04].

4.2.2.3. *The ligaments*

Anterior longitudinal ligament (ALL): the ALL lines the anterior side of the FSU and connects to the vertebral bodies and the IVD (Figure 4.4). Its superficial fibers are stretched over 3–5 vertebrae, while its deep fibers are stretched over two vertebrae [RAN 04]. This ligament provides an anterior passive immobilization of the FSU which restricts extension movement.

Posterior longitudinal ligament (PLL): the PLL is located on the posterior side of the vertebral bodies and covers the anterior wall of the spinal canal (Figure 4.5). Wide at the rear part of the IVD and very narrow in the median part of the vertebral bodies, just like the ALL, it consists of two tendinous layers (Figure 4.5).

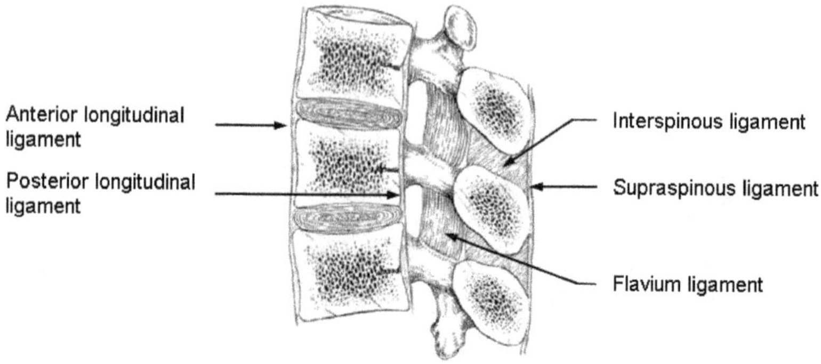

Figure 4.4. *Ligaments of the spine*

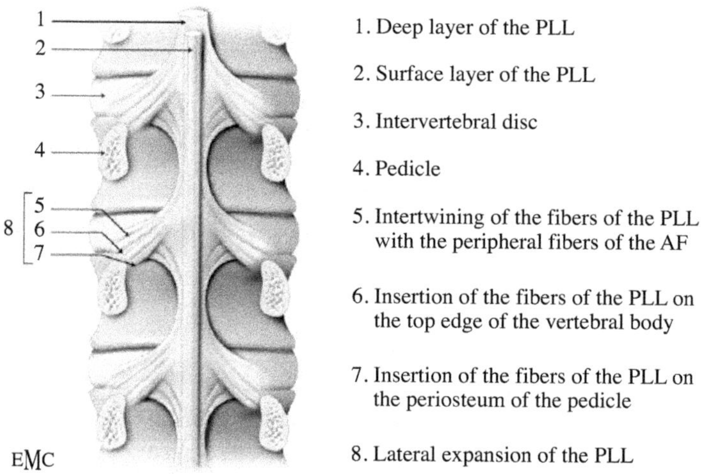

Figure 4.5. *Dorsal view of the posterior longitudinal ligament (PLL) after section of the pedicles*

Posterior ligaments: in the posterior part of the FSU, we find ligaments: ligamentum flavum (LF), intertransverse, interspinous and supraspinous which, respectively, unite the laminar, traverse processes and spinous processes. The LF is the ligament richest in elastin. It is very thick (up to 1 cm) and has two layers [VIT 07]. The interspinous ligaments offer weak resistance to the bending movements related to the orientation of the collagen fibers. In fact, these ligaments are in continuity with the thoracolumbar fascia muscle that they tie firmly [RAN 04]. The supraspinous ligaments, which are loose and poor in collagen, participate in an insignificant manner in stabilization. However, they avoid traumatic lesions of spinous processes, the insertion area of several spinal muscles [RAN 04].

4.2.3. *Intervertebral disk prosthesis*

Artificial disks are implanted from an anterior approach, which has the advantage of leaving the posterior structures intact and largely responsible for spinal stability. They usually consist of two metal plates anchored to the surface of the vertebral bodies and a convex-shaped polymer or metal part which is joined to the concave metal endplates (Figure 4.6). Today, several artificial disks are available on the market, but only a few of them have been approved by the Food and Drug Administration (FDA) such as Charité III (DePuy Spine) and ProDisc (Synthes) [SER 11].

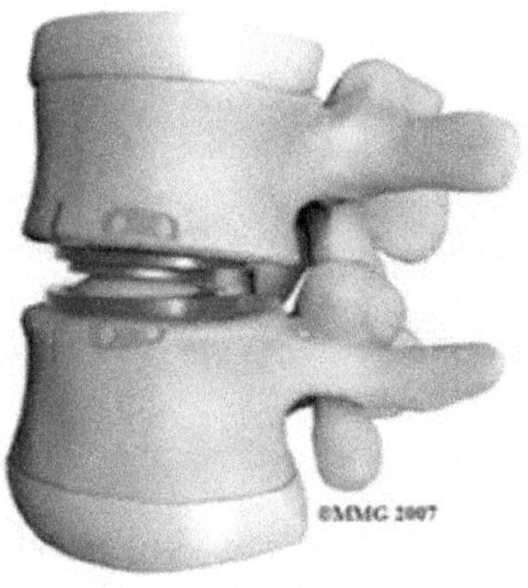

Figure 4.6. *Artificial disk in situ*

The Charité III artificial disk is the third version of the first marketed modern artificial disk. It was certified by the FDA in 2004. The concept of this implant is based on hip and knee prostheses which mostly use combinations of polyethylene-metal materials for joint surfaces. The current version has a polyethylene mobile core (UHMWPE) which presents articulated convex surfaces with concave endplates made from CoCrMo alloy covered by a titanium coating and a calcium phosphate layer (Figure 4.7(a)). The polyethylene mobile core allows a sliding movement and a translation of instantaneous center of rotation following the anteroposterior axis during flexion-extension movements. Serhan *et al.* [SER 11] quote Cunningham's *in vitro* studies which show that the Charité III prosthesis restores natural mobility of the segment in flexion-extension and lateral flexion but gives higher than normal mobility in axial rotation. Moreover, the kinematics of the adjacent level is not modified. Serhan *et al.* [SER 11] also note that clinical studies performed in Europe announce an average intervertebral flexion-extension mobility of 10.3° of the segment after 10 years of follow-up. However, some clinical failures have been reported recently. Unlike Charité III, the ProDisc prosthesis has a fixed center of rotation because the polyethylene core is fixed to the lower endplate of the implant (Figure 4.7(b)). The studies show an average mobility range of 3.8° (radiographic measurement) after more than 8 years of clinical follow-up [SER 11].

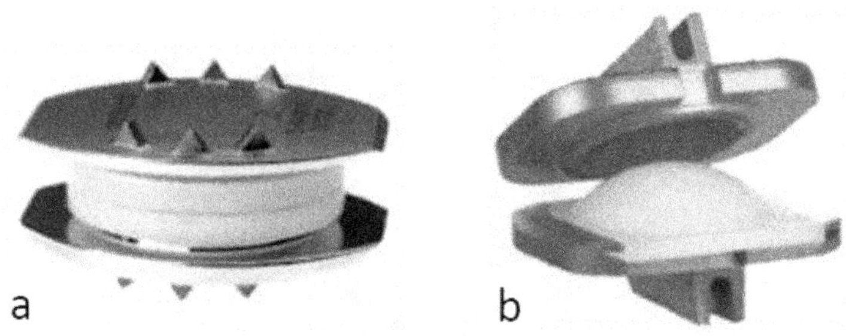

Figure 4.7. *Disk prostheses approved by the FDA: a) Charité III and b) ProDisc*

Artificial disks can be used on one or two vertebral levels. They are aimed at treating patients suffering from disk degeneration with a disk height of at least 4 mm and with no articular surface degeneration. However, their use remains very limited because there are a number of contraindications (deformations, spondylolisthesis, osteoporosis, etc.) [KAN 10]. The clinical results are heavily dependent on the selection of patients [LAU 07]. Lemaire *et al.* [LEM 07] analyzed the complications of lumbar disk prostheses on a series of 200 cases. They conclude that the complication rate of these prostheses is equivalent, or even lower than that of anterior arthrodeses and that the short- and long-term rate of revision surgeries is markedly lower. The disadvantages of these devices continue to be related to low shock absorption capability, limited indications and migration risk.

4.2.4. *Numerical application on the intervertebral disk*

The purpose of this application is to carry out a simple reliability analysis using commercial software such as ANSYS. Figure 4.8 shows an intervertebral disk located between the lumbar vertebrae $L4$ and $L5$.

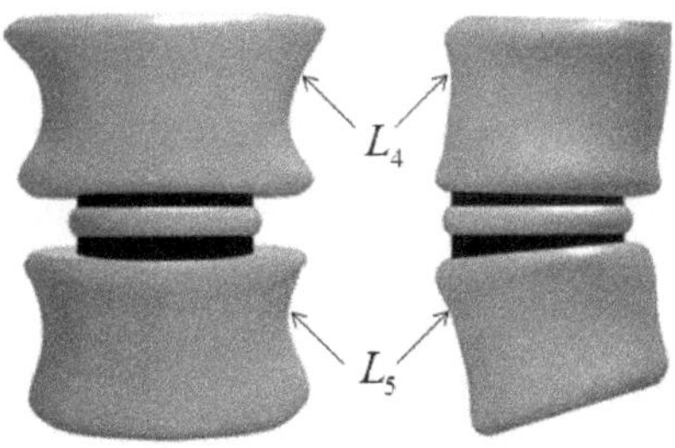

Figure 4.8. *Position of the studied disk between lumbars L4 and L5*

The dimensions of the studied two-dimensional (2D) geometrical model are: $D1 = 50\,mm$, $D2 = 46\,mm$, $D3 = 40\,mm$ and $H = 10\,mm$ (Figure 4.9). The disk is subjected to a pressure ($P = 100\,MPa$) and made up of polyethylene (UHMWPE) with Young's modulus: $E = 100\,GPa$ and Poisson's ratio: υ = 0.3. The yield stress of the material used is $\sigma_y = 75\ MPa$. Considering the global safety factor ($S_f = 1.5$) and design variables ($D1\ \&\ D2$), we minimize the volume under allowable stress conditions. After this, the Hasofer–Lind reliability index is calculated.

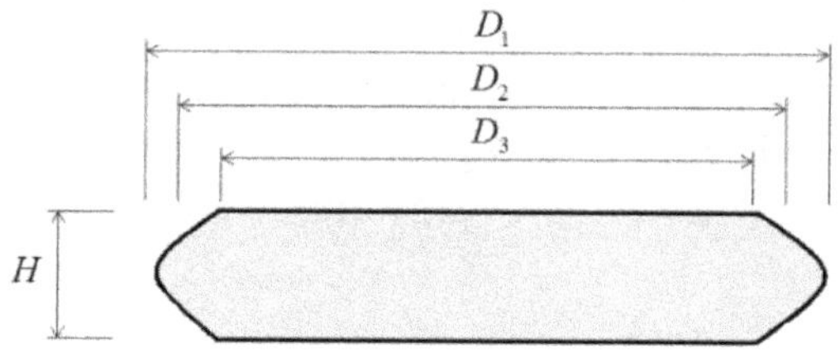

Figure 4.9. *2D model of the studied disk*

Using ANSYS software, the study of this disk can be carried out in three main steps (see code in Appendix 2):

4.2.4.1. *Numerical simulation using finite elements*

This step deals with (Figure 4.10):

– *Input of the initial values of the parameters* ($D1\ \&\ D2$): these parameters should be entered at the beginning as a starting point for optimization and later to allow relating them with the geometrical elements (points, lines, etc.).

– *Geometrical model construction*: the geometry model can be created from the basic elements (points, lines, areas and volumes). However, it is possible to use existing elements in the software (rectangle, circle, etc.). For the studied model, we have created the model from points and lines. Next, we build the areas (surfaces) from these lines (Figure 4.11).

– *Meshing model creation*: the total number of elements considered for the model is 54 (4-node/PLANE42) and the number of nodes is 72 (Figure 4.12).

– *Introduction of boundary conditions*: the disk is subjected to pressure (100) and fixed from the bottom (Figure 4.13).

– *Solving of equation system*: depending on the built models (geometry and meshing) and the provided boundary conditions, the stiffness matrix is created to solve the static equations system.

– *Handling of results*: Figure 4.14 shows the distribution of the von-Mises stresses of the direct simulation by finite elements. The initial volume is 461.43 mm^3 and the maximum von-Mises stress is $\sigma_{max} = 10.25$ MPa.

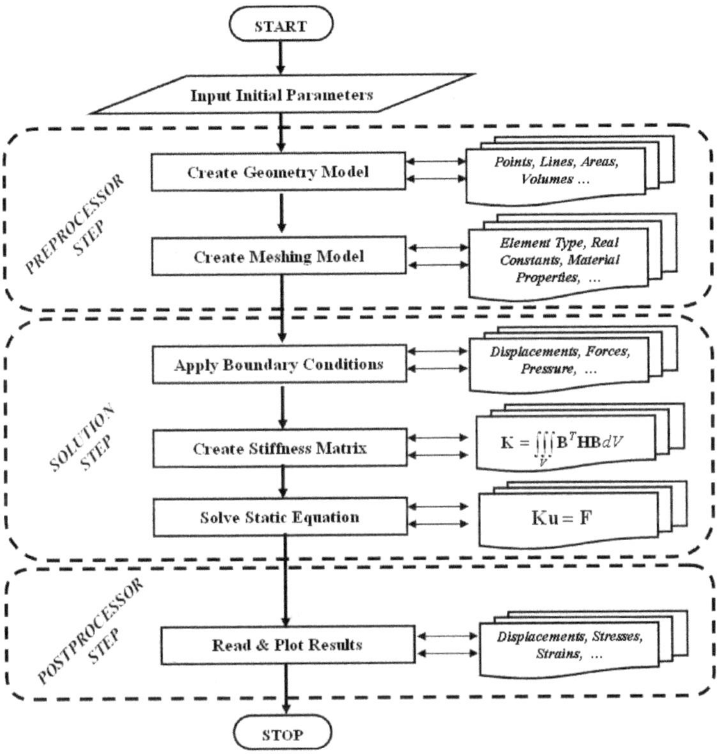

Figure 4.10. *Disk simulation strategy*

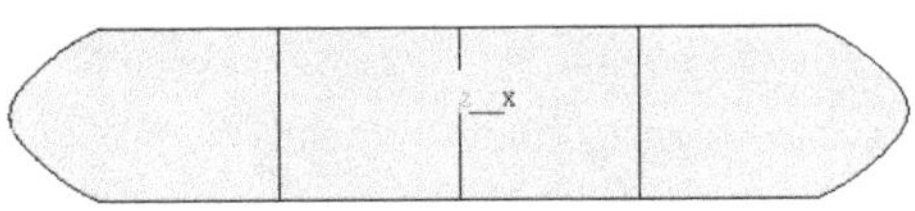

Figure 4.11. *Transformation of the geometry model into areas (surfaces)*

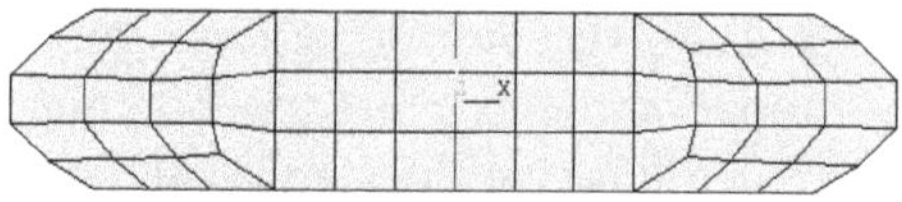

Figure 4.12. *Meshing model*

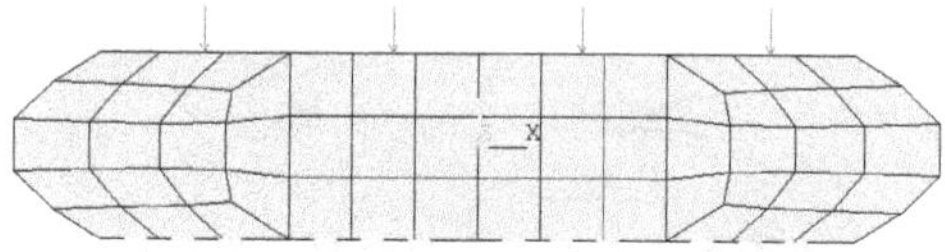

Figure 4.13. *Boundary conditions*

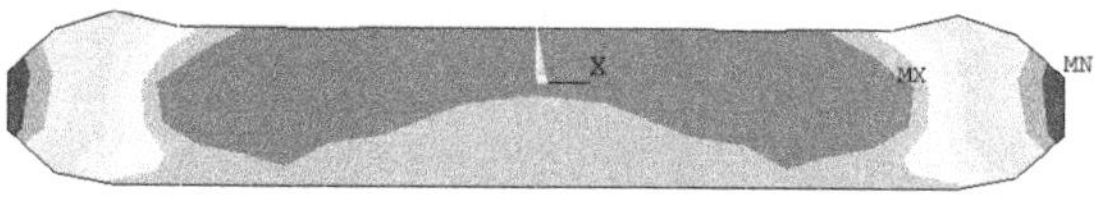

Figure 4.14. *Von-Mises stresses for the starting point (initial)*

Figure 4.10 shows the general numerical simulation approach by finite elements with the use of a finite element software such as ANSYS.

4.2.4.2. *Optimization for the optimal solution*

The optimization problem of two variables seeks to minimize the volume of the disk under the allowable stress condition. It can be written in the following form:

$$\begin{aligned} &\min : Volume \\ &\text{subject to:} \sigma_{\max} - \sigma_w \leq 0 \end{aligned} \qquad [4.1]$$

The allowable stress is thus calculated using the global safety factor:

$$\sigma_w = \frac{\sigma_y}{S_f} = \frac{75}{1.5} = 50\ MPa \qquad [4.2]$$

where σ_y is the yield stress of the disk material. This optimization procedure can be carried out in the ANSYS Mechanical module (APDL) up to ANSYS 13.0 version. However, we can use the commands of the ANSYS Mechanical module (APDL) in the ANSYS-WorkBench module for recent versions (starting from ANSYS 14.0 version). Figure 4.15 shows the 2D geometric configuration of the optimal solution. The optimal volume is 388.31 mm^3 and the maximum von-Mises stress is $\sigma_{max} = 49.98 \approx \sigma_w$ MPa (Table 4.1).

Figure 4.15. *Geometric configuration for the optimal solution*

Parameters	Initial point	Optimal solution	Design point	
			Direct method	Classical method
D1	50	32.614	30.092	30.303
D2	40	41.814	38.021	40.337
Volume	461	388.31	358.24	373.04
σ_{max}	10.68	49.981	74.819	74.000

Table 4.1. *Mechanical results of the studied intervertebral disk*

4.2.4.3. *Calculation of reliability*

When using ANSYS software, there is no module for calculating reliability. A direct method has been developed to simplify the calculation of reliability with two simple optimization loops [KHA 09]. However, in order to implement the classical reliability method, we need to innovate certain techniques using APDL (ANSYS Parametric Design Language) commands.

4.2.4.3.1. Direct method

The idea behind this method is first to find the coordinates of the design point and then to calculate the reliability index by considering the difference between the optimal solution and the design point in the normalized space [KHA 09].

In order to find the design point, an optimization problem of two variables seeking to minimize the volume of the disk under the yield stress condition is defined. It can be written as follows:

$$\begin{aligned}&\min : Volume\\&\text{subject to: } \sigma_{\max} - \sigma_y \leq 0\end{aligned} \qquad [4.3]$$

Figure 4.16 shows the geometric configuration of the design point by the direct method. The resultant volume is: 358.24 mm^3 and the maximum von-Mises stress is $\sigma_{\max} = 74.82 \approx \sigma_y$ MPa (Table 4.1).

Figure 4.16. *Geometric configuration for the design point using the direct method*

Reliability can be calculated by the Hasofer–Lind reliability index method. The reliability index is calculated in the normalized space of the two variables. The standard deviations σ_i are proportional to the mean values ($\sigma_i = 0.1\, m_i$). The normalized variables follow the normal distribution law and are calculated in the following form:

$$u_{D1} = \frac{D1 - m_{D1}}{\sigma_{D1}} = \frac{30.092 - 32.614}{0.1 \times 32.614} = -0.773 \qquad [4.4]$$

$$u_{D2} = \frac{D2 - m_{D2}}{\sigma_{D2}} = \frac{38.021 - 41.814}{0.1 \times 41.814} = -0.907 \qquad [4.5]$$

Then, the reliability index is calculated by:

$$\beta = \sqrt{u_{D1}^2 + u_{D2}^2} = 1.192 \quad [4.6]$$

Using equation [2.85] (or Appendix 4), the probability of failure is therefore:

$$P_f = \Phi(-\beta) = \Phi(-1.192) \approx 0.117 \quad [4.7]$$

The reliability is therefore:

$$R = 1 - P_f = 1 - 0.117 = 88.3\% \quad [4.8]$$

Table 4.1 shows the different results of this method. This method is easy to implement to give a general and simple idea of reliability analysis.

4.2.4.3.2. Classical method

The classical method concerns the minimization of the distance between the limit state function and the origin of the normalized space with the aim of finding the most probable failure point (MPP) point. We can introduce the formulation of the reliability index $\beta = \sqrt{u_{D1}^2 + u_{D2}^2}$ directly as an objective function. The normalized variables $\mathbf{u}$ or the random ones $\mathbf{y}$ can be variables of optimization by taking into account the correlation between themselves (normal distribution law). The maximum von-Mises stress is considered herein as a limit state function. Figure 4.17 shows the geometric configuration of the design point by the classical method. The resultant volume is 358.24 mm^3 and the maximum von-Mises stress is $\sigma_{\max} = 74.82 \approx \sigma_y$ MPa (Table 4.1).

Figure 4.17. *Geometric configuration for the design point using the classical method*

As shown in Table 4.2, the resulting value of the reliability index $\beta = 0.8$ is less than that resulting from the direct method $\beta = 1.19$.

We can note that the use of an additional technique helped in improving the optimal value of the reliability index for a better MPP.

	Direct method	Classical method
β	1.192	0.80
P_f	11.74%	21.24%
R	88.26%	79.76%

Table 4.2. *Probabilistic results of the studied intervertebral disk*

4.3. Reliability analysis of the hip prosthesis

First, we present some information on the hip's anatomy and then we will focus on the hip prosthesis in order to apply the reliability analysis.

4.3.1. *Anatomy*

4.3.1.1. *Different views*

The hip is the region of the human body that unites the torso to the lower limb. It comprises a joint called *coxofemoral joint* (Figure 4.18(a)), muscles which allow it to move, as well as neurovascular bundles which cross it. This coxofemoral joint will be the seat of the main pathologies of this region. It relates the pelvis *(coxal bone)* (Figures 4.18(c) and (d)) to the femur *(femoral head).* This is a robust joint, of very great stability but with a limited range of movements compared to the shoulder joint which is much more mobile but more fragile. Its purpose is to orientate the lower limb in all spatial directions, and is very well adapted to the upright position. The acetabular area (Figure 4.18(b)) receives and totally covers the femoral head.

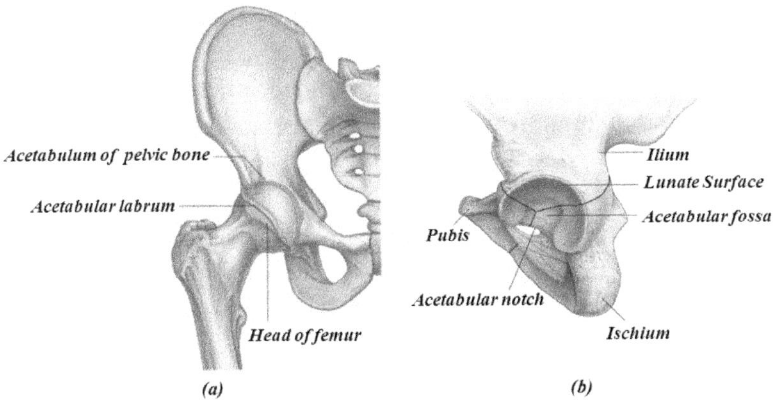

Figure 4.18. *a) Coxofemoral joint, frontal view, and b) Acetabulum in the coxal bone, lateral view*

4.3.1.2. *Articular surfaces of the coxofemoral joint*

4.3.1.2.1. At the femur level

The *femur head* is the proximal end of the femur which joins with the coxal bone (Figures 4.19(a) and (b)). It is a rounded protrusion, forming 2/3rd of a sphere, supported by the femoral neck which ensures the junction with the femoral body, and which looks inward, on top and projecting slightly forward.

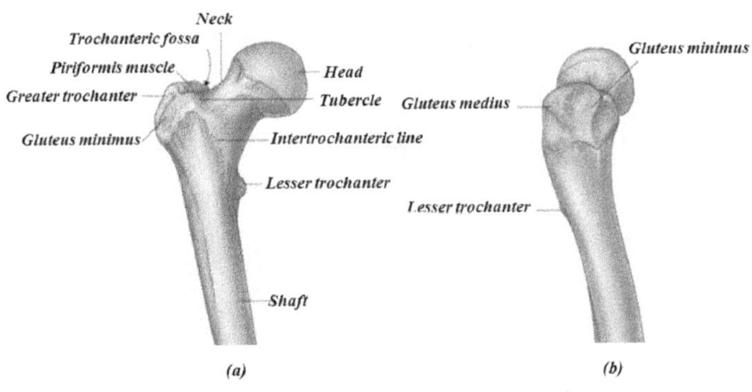

Figure 4.19. *Proximal extremity: a) Frontal view, and b) Lateral view*

The femoral head is lined by a layer of cartilage. At its center, we find the fovea to which a ligament is attached: *the round ligament,* which contributes to restricting combined movements of flexion, adduction and external rotation.

4.3.1.2.2. At the coxal bone level

The articulation with the femur is done by a hollow articular surface in the pelvic bone: *the cotyloid cavity*_(or cotyl or acetabulum) (Figure 4.20). This cavity, which is hemispheric, presents two parts:

– one articular part, crescent-shaped (lunatic surface), with a bulge attached to its rim thereby increasing its contact surface area: *the acetabular labrum*;

– the other one, retracted and non-articular, which is called the *acetabular fossa.*

This articular cavity looks toward the bottom, outward and forward. When these two surfaces come into contact, we can note that in the upright position the femoral head is poorly covered by the cotyl. In fact, surface coincidence takes place when the hip is bent at 90°, which corresponds to a vestige of the quadrupedal position.

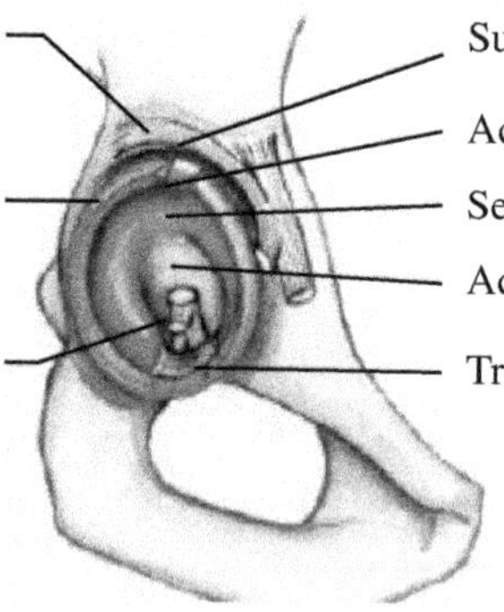

Figure 4.20. *Acetabular anatomy*

4.3.1.3. *Means of union*

The articular surfaces are kept in contact by an articular capsule, a highly solid fibrous sleeve stretching between the coxal bone and the base of the femoral body on the one hand, by anterior and posterior ligaments which reinforce this capsule on the other hand, and by the peri-articular muscles (Figure 4.21). The layout of the orientation of the fibers of this capsule and its ligaments ensures excellent robustness while allowing mobility of this joint.

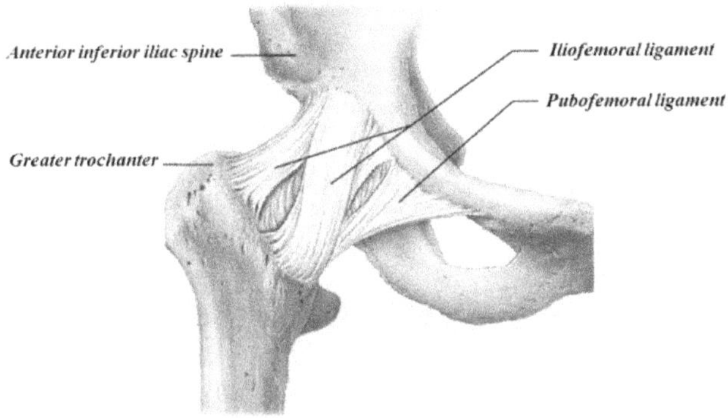

Figure 4.21. *Ilio-femoral and pubo-femoral capsule and ligaments*

4.3.1.4. *Muscles enabling hip mobility*

The hip is less mobile than the shoulder. Bending the hip brings the frontal side of the thigh toward the torso and its amplitude depends on the position of the knee. When the thigh touches the torso, some of the mobility comes from the lumbar spine. The muscles which enable bending are the anterior muscles of the thigh and some muscles from inside the abdomen. Extension of the hip is more limited, rendered possible by the spine muscles of the buttock and thigh. Abduction, which brings the lower limb outward, is primordial for walking and is mainly taken care of by the buttock muscles which thus take care of the equilibrium of the pelvis.

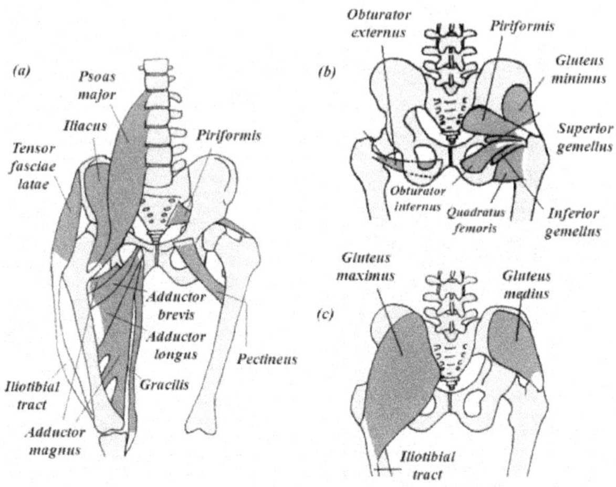

Figure 4.22. *Several hip muscles: a) anterior view, and b) posterior view, and c) Gluteus maximus and Gluteus medius, posterior view*

4.3.2. *Presentation of the total hip prosthesis*

Orthopedic surgery and in particular hip prosthesis surgery are developing continuously. A total of 800,000 THPs are implanted each year worldwide. The exceptional craze of patients and surgeons for this prosthesis comes from the fact that it yields exceptional and tangible results insofar as pain relief, increase in mobility and low duration of surgical outcomes are concerned. Nowadays, with the help of a physiotherapist, patients can get up a few days after the surgery. Arthrosis and degradations of the coxo-femoral joint which mostly affect the aged population have become a major public health issue. The wear and tear of the cartilage, the precious lining which enables harmonious sliding of the femoral head inside the cotyloid cavity leads to malfunction of the joint. Furthermore, this symptom is often accompanied by other complications like intense pain. Medicines are not always effective to treat this condition and a hip arthroplasty often proves to be necessary. Young people are also affected. Their pathology is on the rise owing to new life and work styles, and is also aggravated by certain disastrous sports practices. Many sports, for example, football, basket-ball, skiing or even jogging, put a lot of strain on the hips. In this case, these are sudden movements, directional changes, rapid to and fro movements, blocks at landing after a jump, etc., which make the hips bear loads that could be 10 times the body weight. If there is no satisfactory conservative treatment available, hip prosthesis is generally suggested to patients in order to get back complete mobility and improve their lifestyle. Right at the beginning of implantations, it was thought that the mechanisms responsible for acetabular loosening were mechanical in nature, in relation to friction problems, resistance to wear and tear and design of parts. Subsequently, researchers and practitioners laid emphasis on the role of biological factors: the organism's reactions to the particles produced by the wear and tear of implants. Moreover, all these reactions take place in a physiological liquid which involves the medium's action on the long-term performance of the materials.

Literature in the area of THPs is very rich but studying it can be confusing owing to the extreme diversity of the results. Through this literature review, we will try to describe the characteristics of THPs, identify the causes for failure and describe the consequences thereof.

We thus begin to get an idea of the complex nature of a comprehensive study of the degradation phenomenon of a total hip replacement prosthesis. We have opted to focus on the friction of biomaterials constituting a hip implant, in a physiological medium. After technological advancements made in the tribological domain, the design and manufacturing of THPs has seem considerable changes. Thus, surgical implantation of prostheses has become one of the most common surgical acts. Despite these changes, a number of problems still remain to be resolved. Cracking, and wear and tear of the components of total hip arthroplasties remain as an unavoidable mechanism. It is in this spirit that we will be focusing our work on the various shapes and design which will be the first support elements to improve the performance of prostheses

THPs are different from one another in their architecture (cervico-diaphyseal angle and external lever arm), in the form of the stem (straight or more or less curved, with a rectangular, oval, diamond, cylindrical or a variably decreasing section), by the appearance of the smooth surface, glossy or matte, fine-grained, by the nature of the material used (stainless steel and chromium cobalt, titanium) the mechanical characteristics of which are different.

However, in first-line treatment THPs, the implant should achieve high fatigue strength. In fact, the unipodal stance during walking gives rise to a vertical stress of around 3–4 times the body weight at hip level. Running, jumping and going down the stairs at a brisk pace cause more considerable stresses reaching up to 6–7 times the body weight, thus giving rise to the need for tough material with a geometrical shape likely to remove any concentration of stresses in fragile zones. The cause of detachment is essentially bone resorption for which several factors may be responsible: excess or insufficiency of stresses, or reaction to a foreign body. Stress shielding is the oft-quoted phenomenon to describe the absorption of proximal bone stresses by the implant, which restores them by its distal end.

In general, a THP is made ups of four parts: the acetabular cup, the liner, the femoral head and the femoral stem (Figure 4.23):

– *The acetabular cup*: this is a hollow part, usually metal, which is sealed in the pelvis receiving the prosthetic head. They can be cemented or uncemented with different sealing techniques;

– *The liner*: is usually made up of plastic materials and in contact with the cup;

– *The femoral head*: this is the end of the femoral prosthesis which joins with the cup through the liner to form the prosthetic joint. It can be independent of the stem or otherwise. They are differentiated by their diameter: 22, 28, 32 mm and their materials (friction couple);

– *The femoral stem*: the lower part of the THP is the part which deals with the stresses. It is introduced into the femur after preparing it. It can be sealed in the femur with acrylic cement, this is the case for stems known as "*straight*" stems, or forcefully blocked in the canal of the femur, this is the case for "*anatomical*" stems which fit the morphology of the bone, and which can be uncemented and covered with hydroxyapatite or cemented.

The prosthesis can be *modular*, i.e. it may contain several segments, each of a different size, which fit into one another to make the final implant. The modularity of the femoral part concerns the stem-head junction.

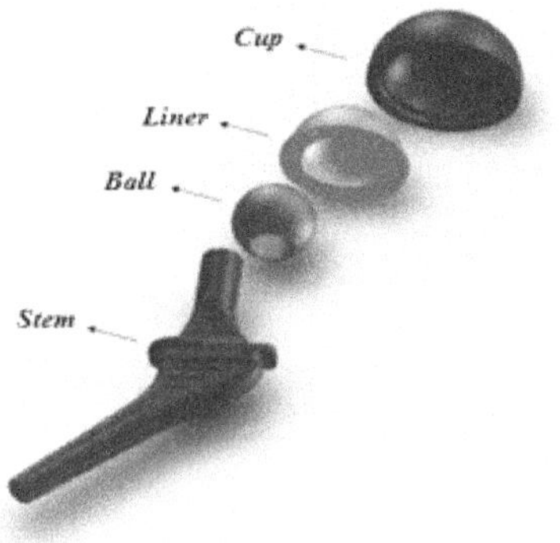

Figure 4.23. *Main parts of a total hip prosthesis*

4.3.3. *Numerical application of the hip prosthesis*

Figure 4.24 shows the metal stem of the hip prosthesis in three-dimensions (3D). The problem with 3D is that it requires a parameterized geometry to be optimized which leads to a significant amount of calculation time [KHA 16a].

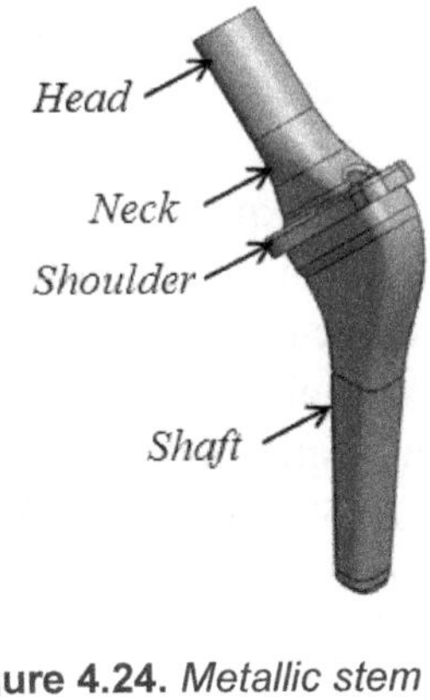

Figure 4.24. *Metallic stem in 3D*

For this reason, we can simplify the model to make it 2D. Figure 4.25 shows a section of the different layers of the femoral part of the hip prosthesis: the metal stem, the cortical bone, the trabecular bone and the bone marrow.

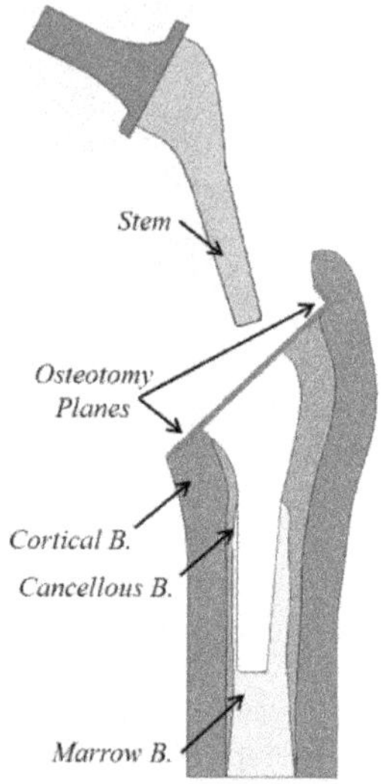

Figure 4.25. *Femoral part of the hip prosthesis with the different bone layers*

The total number of elements considered for the model is 1,476 (8-node/PLANE82) and the number of nodes is 4,825 (Figure 4.26). The cortical bone is considered homogeneous and isotropic with an elasticity modulus (Young's modulus) of $E = 17\,\text{GPa}$ and a Poisson's ratio of $\upsilon = 0.3$. The experimental value of its yield stress is $\sigma_y = 132$ GPa [KEA 01]. The total number of elements considered for this region is 699. The trabecular bone is considered homogeneous and isotropic with an elasticity modulus (Young's modulus) of $E = 761\,\text{GPa}$ and a Poisson's ratio of $\upsilon = 0.3$. The experimental value of its yield stress is $\sigma_y = 10.53$ GPa [ALE 13]. The total number of elements considered for this region is 220. However, the bone marrow layer can be seen as a cavity and therefore neglected at the time of analysis of finite elements [TSU 02]. Titanium alloy is considered to be the most biocompatible material among all prosthetic materials. This is why the femoral stem's material is considered to be made of titanium alloy (Ti-6Al-4V). The properties of the material of the hip's femoral proximal stem are Young's modulus ($E = 110$ GPa), Poisson's ratio (υ = 0.3) and the yield stress (σ_y = 800 MPa) [KAY 08]. The total number of elements considered for this metallic region is 557.

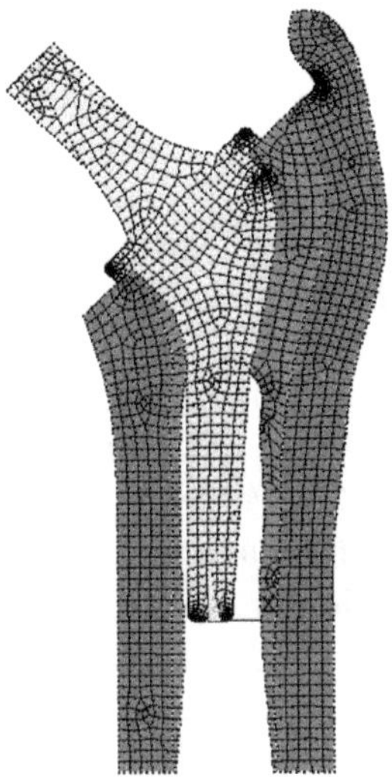

Figure 4.26. *Meshing model. For a color version of the figure, see www.iste.co.uk/kharmanda/biomechanics.zip*

The performance of a hip replacement is directly related to the ability to transfer the load uniformly from the components to the surrounding bone. A region of the bone which is subject to low load levels may be subjected to absorbance (Wolf's law [WOL 86]) which will lead to loosening and eventually loss of functionality of prostheses. For this reason, the minimum stress value in the surrounding bone has to be kept above a certain minimum level. The minimum stress is related to the natural strain energy density (SED). For linear elasticity, the SED can be written as:

$$U = \frac{1}{2} \varepsilon : \mathbf{D} : \varepsilon \qquad [4.9]$$

where **ε** is the strain tensor and **D** is the elasticity tensor. According to Huiskes *et al.* [HUI 87], the natural SED is estimated at the periosteal bone surface as: $U_N = 5.03 \times 10^{-6}$ MPa. On the basis of equation [4.9], the minimum stress value can be computed as follows:

$$\sigma_{Tar} = \sqrt{2EU_N} \qquad [4.10]$$

where E is the uniaxial Young's modulus. To ensure long-term fixation, the minimum stress value in the surrounding bone has to exceed the threshold value σ_{Tar}.

4.3.4. *Boundary conditions*

Three representative daily loading conditions of one-legged stance (L1), extreme ranges of motion of abduction (L2) and adduction (L3) are considered [KHA 15a], as shown in Figure 4.27. The boundary conditions at the distal end have no effect on the stresses in the proximal region. The fixation is carried out on lower bone cut (on the cortical layer) to avoid rigid-body motion.

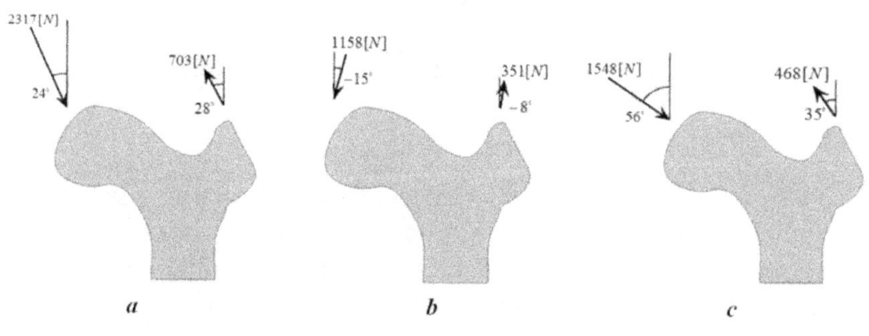

Figure 4.27. *Boundary conditions: L1, L2 and L3 [KHA 16a]*

4.3.5. *Direct simulation*

The output parameters can be represented by an indication of fracture stresses and loosening of prosthesis. The von-Mises stresses give an indication of the fracture stress at the different layers of the studied structure. The minimum stress value in the surrounding bone has to be kept above a certain minimum level (equation [4.10]) to avoid the loosening of prosthesis functionality. According to Table 4.4, the third loading case is the most critical one. The limit state function is represented by the trabecular region stresses. Here, we note that the maximum von-Mises stress ($\sigma_{\max}^{2} = 5.63$ MPa).

Materials	E [MPa]	ν	σ_y [MPa]	σ_T [MPa]	Number of elements
Ti-6Al-4V Alloy	110,000	0.3	825	–	557
Cortical	17,000	0.3	132	0.41	699
Trabecular	761	0.3	10.53	0.09	220

Table 4.3. *Mechanical properties of the used materials*

Output parameters	L1	L2	L3
$\sigma_{\max}^{1}$ (MPa)	12.94	9.02	15.59
$\sigma_{\max}^{2}$ (MPa)	3.90	3.46	5.63
$\sigma_{\max}^{M}$ (MPa)	16.75	12.23	11.28
$\sigma_{\min}^{M/B}$ (MPa)	0.21	0.33	0.34

Table 4.4. *Output parameters for a direct simulation*

Figure 4.28 shows the von-Mises stress distribution for the three loading cases. In the first and second cases, the maximum stresses are located in the metallic stem.

However, in the third one, the maximum von-Mises stress is located at the bottom right region of the cortical layer. Here, we can distinguish a tension failure case (Table 4.4).

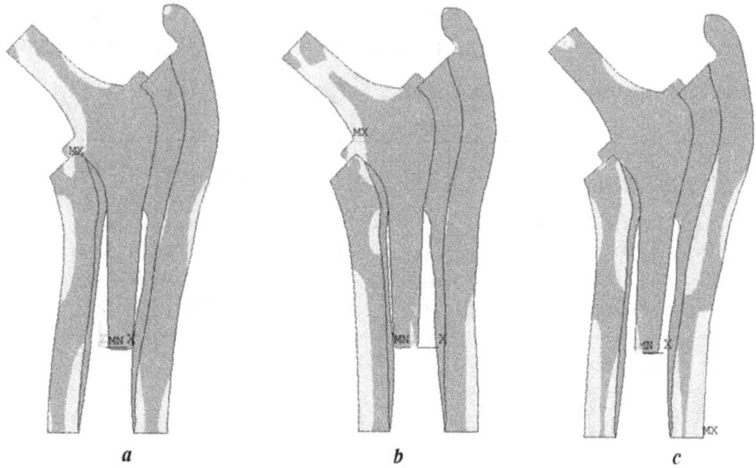

Figure 4.28. *Results of the initial simulation. For a color version of the figure, see www.iste.co.uk/kharmanda/biomechanics.zip*

4.3.6. *Probabilistic sensitivity analysis*

The evaluation of probabilistic sensitivities is based on the correlation coefficients between all random input variables and a particular random output parameter. The sensitivity plots only include the significant random input variables. Figure 4.29 shows the sensitivity measurements of the limit state function (output parameter) with respect to the input random variables (six variables). When considering the three daily loading cases, there is a variant influence of the input parameters. However, the *Young's modulus of the trabecular and cortical regions* has a higher influence relative to the other input parameters. This way when decreasing the Young's modulus of the cortical bone, the maximum stress of trabecular regions will decrease (positive influence). In contrast, when decreasing the Young's modulus of the trabecular bone, the maximum stress of trabecular regions will increase (negative influence).

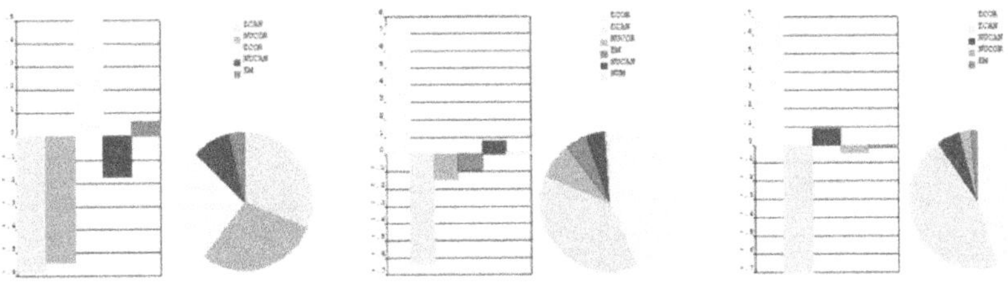

a *b* *c*

Figure 4.29. *Probabilistic sensitivity analysis: a) L1; b) L2 and c) L3. For a color version of the figure, see www.iste.co.uk/kharmanda/biomechanics.zip*

4.3.7. *Integration of reliability analysis*

The objective is to find the MPP which is represented by the minimum distance between the origin of the normalized space and the most critical failure surface (limit state function). According to the previous stem-bone simulation, the limit state function is represented by the von-Mises stress at the trabecular tissue. In order to formulate the reliability problem, equations [3.31] and [4.10] can be integrated to problem [2.79]. Thus, for the given failure scenario (trabecular tissue), the reliability index β is obtained by solving a constrained minimization problem:

$$\begin{aligned}
&\min : d(u_i) = \sqrt{\sum_{i=1}^{n} u_i^2} \\
&s.t. \quad : H(u_i, y_i) = \sigma_{\max}^{2}(u_i, y_i) - R_{T/C}.A_\sigma \left(\frac{y_2}{A_E} \right)^{R_{\sigma/E}} = 0 \\
&\quad : g_1(u_i, y_i) = \sigma_{\max}^{1}(u_i, y_i) - R_{T/C}.A_\sigma \left(\frac{y_1}{A_E} \right)^{R_{\sigma/E}} \leq 0 \\
&\quad : g_2(u_i, y_i) = \sigma_{\max}^{M}(u_i, y_i) - \sigma_y^{M} \leq 0 \\
&\quad : g_3(u_i, y_i) = \sigma_{\max}^{M/B}(u_i, y_i) - \sqrt{2 y_2 U} \leq 0
\end{aligned} \qquad [4.11]$$

Following Hasofer and Lind [HAS 74], it is preferable to work in a standard normalized space of independent Gaussian variables (Figure 2.31(b)) rather than in the space of physical variables

(Figure 2.31(a)). Hence, we adopt the law for a normal distribution, and define a normalized variable u_i by the transformation:

$$u_i = \frac{y_i - m_{y_i}}{\sigma_{y_i}} \qquad [4.12]$$

where y_i is a random variable with the mean value m_{y_i} and standard-deviation σ_{y_i}. The mean value m_{y_i} may be adopted as a design variable x_i. The standard deviations σ_{y_i} are proposed proportional to the mean values ($\sigma_{y_i} = 0.1\, m_{y_i}$). According to the sensitivity analysis, we find two random variables are the most effective in the structure. Here, the physical space and normalized one in a pedagogical manner (bidimensional space) can be modeled in order to get the global optimum. However, when considering several random variables, the results are subject to classical difficulties in nonlinear programming: existence of local minima, gradient approximation and computational time. Since the reliability analysis is carried out in a normalized space, a special technique is developed in order to take advantage of the particular form of the reliability problem using APDL (ANSYS Parametric Design Language). The optimization algorithm, which is illustrated in Figure 4.30, supplies us all the information about the objective and constraint functions. This algorithm minimizes the minimum distance $d(u_i)$ between the origin of the normalized space and the most critical limit state. This procedure is carried out in the normalized space. In the following section, we will study two cases: the pedagogical case of two parameters and the general case of six parameters.

4.3.7.1. *Case 1: Two parameters*

For simplicity, the random variables x_i corresponding to the Young's modulus of the cortical and trabecular bone (E_1, E_2) are normally considered distributed. Their mean values are presented in Table 4.3 and their standard deviations are proposed proportional to the mean values (10%). Table 4.5 shows the reliability indices for three different loading cases when considering two parameters.

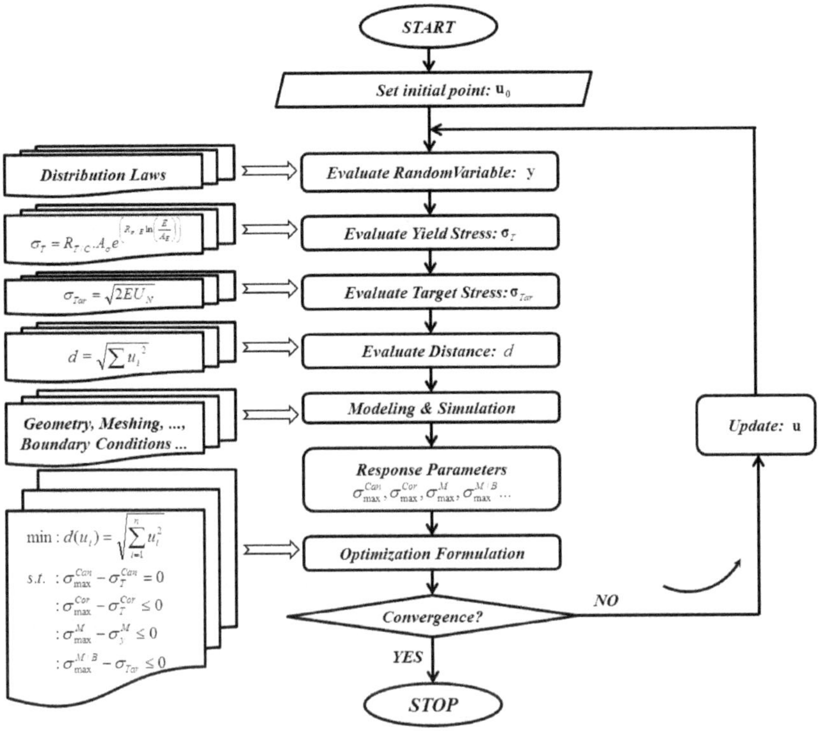

Figure 4.30. *Reliability algorithm*

Figure 4.31 shows the modeling of the optimization problem in the physical space wherein the limit state functions are presented by $G(E_1, E_2) = 0$. However, Figure 4.32 shows the modeling of the problem in the normalized space wherein the limit state functions are presented by $H(u_1, u_2) = 0$. In the physical space, the mean value is presented by the coordinates (x_1, x_2), the MPP is represented by the coordinates (y_1, y_2) and the iso-values of the reliability index are presented by ellipses. The limit states of yield stress in tension are modeled by three curves (full lines: $G_{L1}^T(E_1, E_2) = 0$, $G_{L2}^T(E_1, E_2) = 0$ and $G_{L3}^T(E_1, E_2) = 0$) corresponding to the three loading cases, respectively. The limit states of yield stress in compression are modeled by three curves (broken lines: $G_{L1}^C(E_1, E_2) = 0$,

$G^C_{L2}(E_1,E_2)=0$ and $G^C_{L3}(E_1,E_2)=0$) corresponding to the three loading cases. In this case, the MPP is located on the minimal distance between the mean value and the limit state L3 for the curve of the limit state of yield stress in tension ($G^T_{L3}(E_1,E_2)=0$). However, in the normalized space, the mean value is presented by the initial value (0.0), the MPP is represented by the coordinates (u_1,u_2) and the reliability levels are presented by circles as per equation [2.79]. The limit states of yield stress in tension are modeled using three curves (full lines: $H^T_{L1}(E_1,E_2)=0$, $H^T_{L2}(E_1,E_2)=0$ and $H^T_{L3}(E_1,E_2)=0$) corresponding to the three loading cases. The limit states of yield stress in compression are modeled by three curves (broken lines: $H^C_{L1}(E_1,E_2)=0$, $H^C_{L2}(E_1,E_2)=0$ and $H^C_{L3}(E_1,E_2)=0$) corresponding to the three loading cases. In this case, the MPP is located on the minimal distance between the initial value and the limit state L3 for the limit state curve ($H^T_{L3}(E_1,E_2)=0$). We can also note that the third loading case L3 is the most critical one. The limit state curve divides the space into two regions: safety and failure.

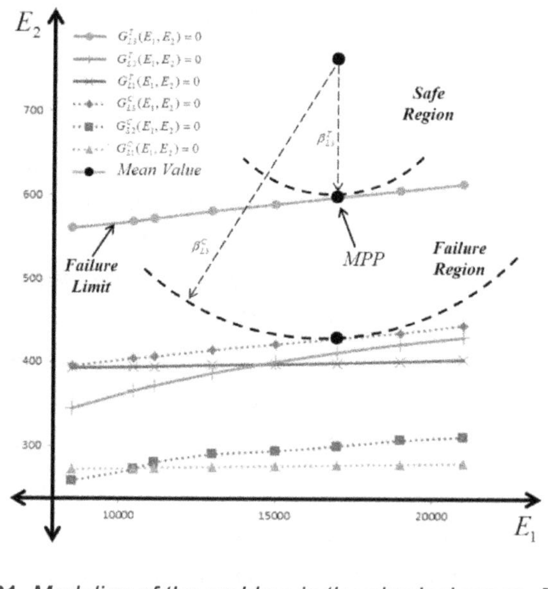

Figure 4.31. *Modeling of the problem in the physical space. For a color version of the figure, see www.iste.co.uk/kharmanda/biomechanics.zip*

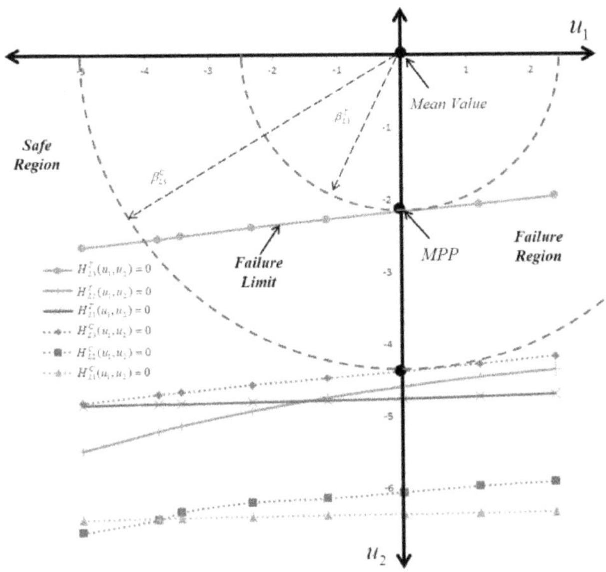

Figure 4.32. *Problem modeling in the normalized space. For a color version of the figure, see www.iste.co.uk/kharmanda/biomechanics.zip*

The MPP is then found for $\beta_{L3}^{T} = 2.16$ when considering the tension failure case that leads to a reasonable level of probability of failure: $P_f = 1.54\%$ using equations [2.86] and [2.87]. However, the reliability index is bigger when considering compression failure case: $\beta_{L3}^{C} = 4.38$ that leads to a very small probability of failure: $P_f = 5.93 \times 10^{-6}$.

Parameters		E_1 (MPa)	E_2 (MPa)	σ_{max}^2 (MPa)	β
L1	$R_{T/C} = 1$	17,000	276.5	4.01	6.36
	$R_{T/C} = 0.7$		398.8	3.96	4.76
L2	$R_{T/C} = 1$		300.7	4.34	6.06
	$R_{T/C} = 0.7$		411.45	4.08	4.59
L3	$R_{T/C} = 1$		427.61	6.04	4.38
	$R_{T/C} = 0.7$		596.5	5.78	2.16

Table 4.5. *Results of the two parameters case*

4.3.7.2. *Case 2: six parameters*

In this case, the random variables x_i corresponding to the Young's modulus and the Poisson's ratio of different layers ($E_1, E_2, E_3, \nu_1, \nu_2, \nu_3$) are normally considered distributed. Their mean values are presented in Table 4.3 and their standard deviations are proposed proportional to the mean values (10%). Table 4.6 presents the reliability indices for three different loading cases when considering six parameters.

For this six parameters optimization process, the MPP is also found for $\beta = 2.16$ when considering the tension failure case that leads to a reasonable level of probability of failure: $P_f = 1.54\%$. The reliability index is higher when considering compression failure case: $\beta_{L3}^{C} = 5.28$ that leads to a small probability of failure: $P_f = 6.46 \times 10^{-8}$.

Parameters	L1		L2		L3	
$R_{T/C}$	1	0.7	1	0.7	1	0.7
E_1 (MPa)	15,747	16,666	16,897	14,055	12,788	16,907
E_2 (MPa)	275.63	399.47	301.32	394.66	412.91	596.97
E_3 (MPa)	110,900	119,850	109,450	107,570	101,490	110,260
ν_1	0.31	0.30	0.30	0.32	0.30	0.30
ν_2	0.31	0.30	0.30	0.32	0.30	0.30
ν_3	0.34	0.30	0.31	0.34	0.31	0.30
σ_{max}^2 (MPa)	4.00	3.96	4.35	3.92	5.84	5.78
β	6.59	4.84	6.05	5.35	5.28	2.16

Table 4.6. *Results of the six parameters case*

Table 4.6 shows that the most efficient parameter is the Young's modulus of the trabecular layer. According to experimental tests by Aleixo *et al.* [ALE 13], the stress/compression rate is $R_{T/C} \approx 1$ (Table 4.6). The experimental results are subjected to various errors (test protocols). In order to improve our design, the value of the rate of $R_{T/C} \approx 0.7$ was considered during the optimization process.

Furthermore, bone composition can be changed as per several factors such as age, disease, food, etc. It is strongly recommended to integrate uncertainty on the behavior of materials during the prosthesis design process. In the literature, several works correlate the mechanical properties of bone materials with its composition [DOB 04].

4.4. Conclusion

The introduction of reliability with respect to orthopedic prostheses seeks to test stability and identify the most critical failure scenario. In this chapter, an introduction of orthopedic prostheses (history and importance) was presented. Two types of prostheses were then studied in a detailed manner: intervertebral disk prosthesis and hip prosthesis. An introduction of the anatomy of each prosthesis was presented followed by detailed numerical applications. The first application on the artificial intervertebral disk is pedagogical in nature and is intended to help understand the possibility of integrating the concept of reliability in a simple way. However, the second application on the hip prosthesis has been presented in a more complex manner along with pedagogical illustration of the calculation of the reliability index in the case of several scenarios of failure.

5

Reliability Analysis of Orthodontic Prostheses

5.1. Introduction to orthodontic prostheses

Orthodontic prostheses can be used to replace teeth (dental implants, bridges, crowns, etc.) or reconstruct teeth (retraction, etc.) or fractures (fixation systems). This chapter focuses on fixation systems for mandibular fractures using mini-plates. It is very important to assess the stabilization of mandibular fractures because this affects the mandible's healing. To do this, a reliability analysis is used to evaluate the safety levels. The effect of the modeling choices on the results has considerable consequences. Taking the anisotropy of the bone into account seems to be essential to mandibular studies. The bone can be considered locally as an orthotropic material that has two orthogonal planes of material symmetry. In a reference point connected to these two planes, its elastic behavior is defined using nine elastic constants or moduli. However, these reference points can change from one point to another in an anatomical part like the mandible by positioning itself in such a way as to guarantee maximal mechanical performances in relation to the usual loads to which it is subject. A change to these loads due to the installation of a prosthesis can lead to a positioning of this orthotropic point of reference. This chapter presents an introduction to the anatomy of the temporomandibular joint. Then, a numerical simulation of a non-fractured mandible is presented by considering two cases of boundary conditions. Finally, two applications of the reliability analysis are carried out to assess the stabilization of the fixation systems used for mandible fractures by considering two cases of the behavior of the bone material (isotropic and orthotropic).

5.2. Anatomy of the temporomandibular joint

Temporomandibular joint (TMJ) dysfunction is essentially related to the meniscal ligament and muscular structure. It can happen that the articular surfaces and, consequently, the joint itself, are partially destroyed. The surgical technique chosen can be the implantation of a prosthesis. The arthrology of the head comes down to the study of the temporomandibular joint (Figure 5.1).

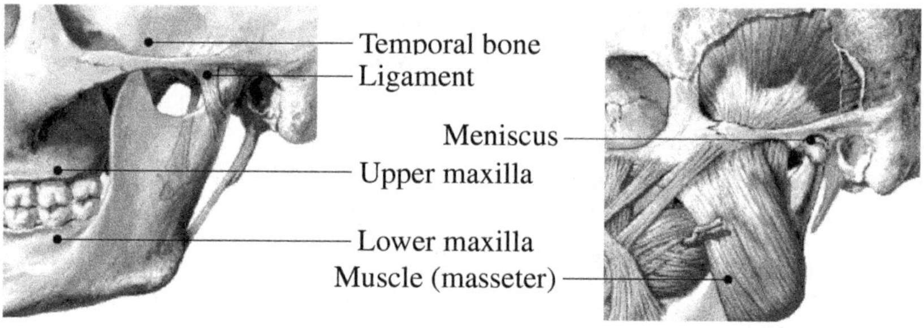

Figure 5.1. *Temporomandibular joint*

The function of the temporomandibular system involves three elementary connections: the two cranial-mandibular joints and the dental occlusion [CAI 00]. The dental occlusion (very intermittently) influences the terminal movements of the mandible and, in this phase, manages the two condyle/glenoid contacts (Figure 5.2) [KUB 88].

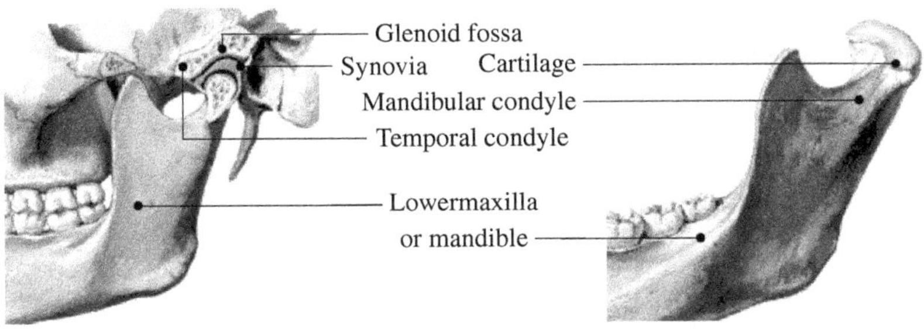

Figure 5.2. *Articular surfaces*

5.2.1. *Articular bone regions and meniscus*

Two bone regions are involved in the basic lateral connection: the skull (temporal zone) and the lower maxilla (mandible). A meniscus is located between them to complete the articular interface.

5.2.1.1. *Bone structure*

5.2.1.1.1. Temporal zone

The temporal surface presents a cavity (glenoid fossa) bordered at the front by the temporal condyle, at the back by the auditory meatus, on the inside by the sphenoid spine and on the outside by the root of the zygomatic process (Figure 5.3). Excluding a pathological situation, the articular surface (temporal slope) meets near the vertex of the condyle.

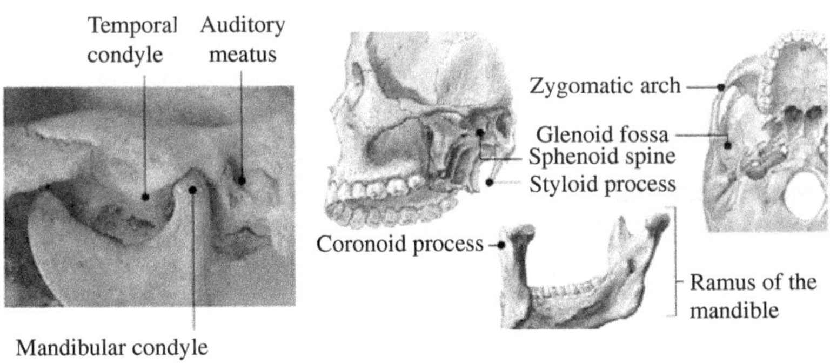

Figure 5.3. *Temporal and mandibular articular bone regions*

5.2.1.1.2. Mandibular zone

The mandibular condyle and the coronoid process (Figure 5.3) make up the tip of the ramus of the mandible. This section, locally comparable to a "long" bone, has a medullary cavity [MOD 89, BIL 00]. The active articular surface is limited to the anterior zone of the condyle.

5.2.1.2. *Meniscus and the articular capsules*

The fibrocartilaginous meniscus can be compared to a biconcave lens that supports the congruence of articular surfaces (Figure 5.4). It has two ridges, one at the front and the other at the back of the smaller area. It is extended by two articular capsules. The temporal capsule connects to the temporal surface by long, loose fibers. The mandibular capsule connects to the condyle surface behind the cartilage and is a taut fibrous cone.

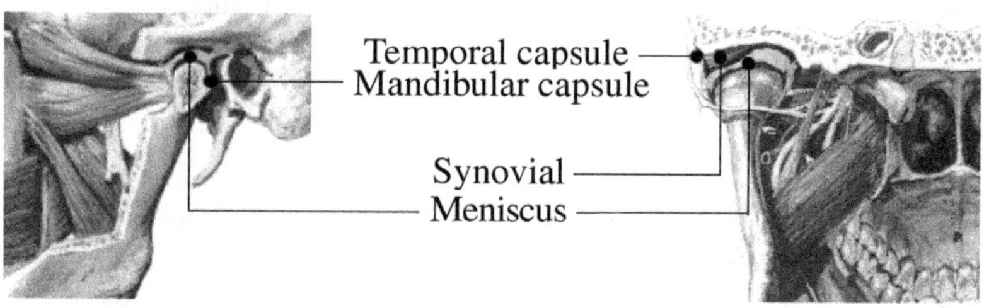

Figure 5.4. *Meniscus, articular and synovial capsules*

Meniscal "brakes" ensure the maintenance and position of the meniscus during temporomandibular movements [CAI 02]. The anterior brake is made up of regions that are not very taut in front of articular capsules. The lateral brakes are composed of ligaments that allow for the sagittal movements of the meniscus (Figure 5.5). The posterior temporal brake is constituted by a very loose temporal capsule and very elastic ligament fibers (Figure 5.4). The posterior condyle brake is completed by a part of the mandibular capsule that is not very elastic.

This anatomical description makes it possible to compare the meniscus to an interface that, with the synovial, reduces the effects of friction on the contact points.

5.2.2. *Ligaments*

Two secondary ligaments and one main ligament, the external lateral ligament, are involved. The external lateral ligament connects to the zygomatic on one side and to the posterior side of the condylar neck on the other (Figure 5.5). The anterior bundle forms a "cord" that is not very elastic and is located in front of the temporomandibular contact area. The posterior bundle and the secondary ligaments reinforce the capsule's action [CAI 02].

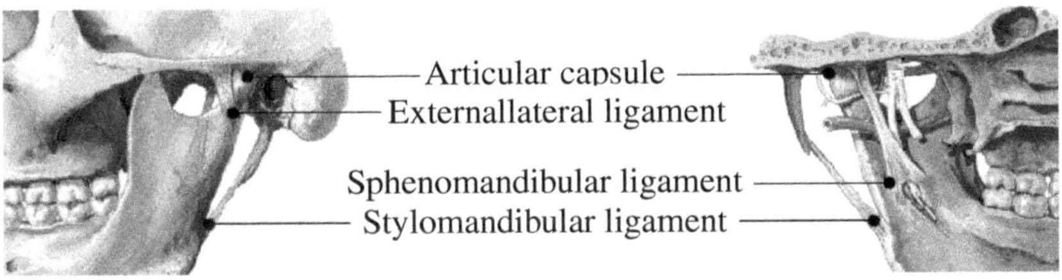

Figure 5.5. *Secondary ligaments and the external lateral ligament*

5.2.3. *Myology, elevator muscles and depressor muscles*

A set of muscles which can be quite powerful, elevators and depressors contribute to the movement of the mandible and transmit actions during the phases of mastication, deglutition or simply speaking [HAN 94].

5.2.3.1. *Elevator muscles*

Three main muscles lift the mandible and contribute to the transmission of actions used for speech and mastication.

5.2.3.1.1. Masseter

A short and thick muscle, the masseter has two main bundles stretched between the zygomatic bone and the zygomatic arch at the lower edge of the angle of the mandible (Figure 5.6). The superficial head (noted as sm), which is external and larger, is a powerful elevator (mastication) of the mandible. Its angle supports the movement of propulsion. The fibers of the deep masseter (noted as dm) maintain a direction close to the normal of the occlusal plane.

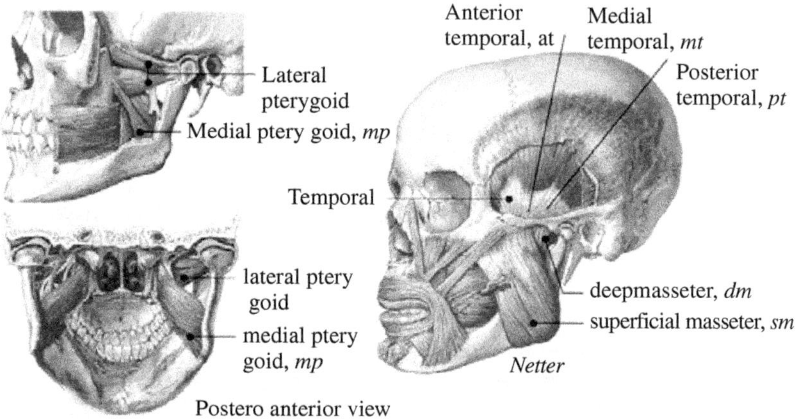

Figure 5.6. *Elevator muscles: masseter, pterygoid and temporal*

5.2.3.1.2. Medial pterygoid

The medial or internal pterygoid muscle (noted as mp) remains parallel to the internal side of the mandible (Figure 5.6). Stretched between the pterygoid fossa and the angle of the maxilla, it is a powerful elevator and is also involved in the movement of propulsion.

5.2.3.1.3. Temporal

The temporal muscle stretches between the coronoid process of the mandible and the temporal insertion surface (Figure 5.6). Large and flat, it can be subdivided into three bundles. The anterior temporal (noted as AT) is both propulsive and an elevator muscle. The middle or medial temporal (noted as MT), whose fibers are almost orthogonal on the occlusal plane, is essentially an elevator muscle. Finally, the fibers of the posterior temporal (noted as PT), which also make up some insertions on the disk, are primarily involved in retropulsive movements and at the end of the lifting.

5.2.3.2. *Depressor muscles*

Three main suprahyoid muscles lower the mandible. Their action is completed by that of the infrahyoid muscles that move the hyoid bone and so provide a point of reference for the suprahyoids.

5.2.3.2.1. Digastricus

The digastric muscle is made up of two bundles connected by an intermediate tendon. The posterior body, stretching from the mastoid process, supports the posterior movement of the hyoid bone (Figure 5.7). The anterior body, stretched between the hyoid bone and the lower maxilla, generates the opening.

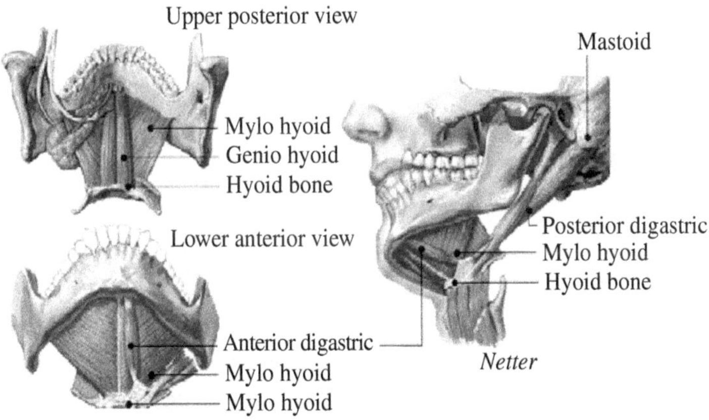

Figure 5.7. *Depressor muscles, hyoid bone and mastoid process*

5.2.3.2.2. Mylohyoid

The large and flat mylohyoid muscle stretches between the lower maxilla and the hyoid bone. It delimits the lower anterior part of the mouth and because of this it is the veritable "floor" of the mouth (Figure 5.7).

5.2.3.2.3. Geniohyoid

The long, narrow geniohyoid muscle stretches between the process of the lower maxilla and the anterior face of the hyoid bone (Figure 5.7).

5.2.3.3. *Lateral pterygoid, unclassified*

The lateral or external pterygoid muscle is divided into two heads (Figure 5.8). The superior head, stretching between the infratemporal crest and the mandibular meniscal-condylar connection, maintains a direction passing through the center of the condyle [HEY 95]. It ensures the position (anterior posterior mobilization) and the stabilization of the disk during mandibular movements. It contributes to the function of the ATM but does not exercise significant action on the mandible during mastication. The inferior head connects the pterygoid process to the mandibular condyle. It is also involved in supporting the meniscus [BER 95].

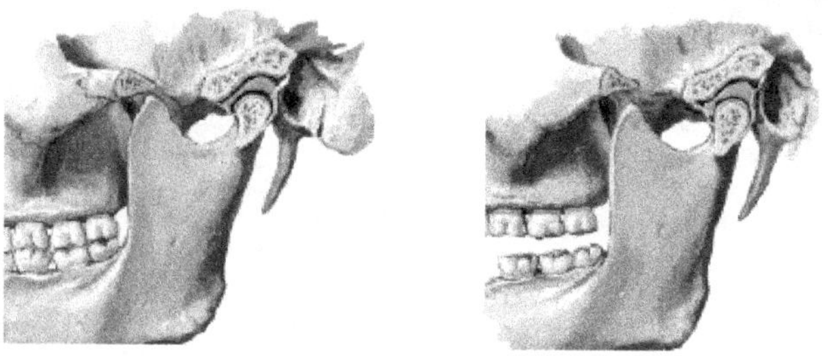

Centric occlusion (intercuspation)

Malocclusion (absence of dental contact)

Figure 5.8. *Centric occlusion, rest position or malocclusion*

The unilateral contraction of the external pterygoid supports the movement of diduction. When the contraction is bilateral and symmetrical, it becomes the propulsive force of the mandible. The internal maxillary artery, which passes between the two heads of the muscle, is constricted during the depressing movement. Because this area is very well irrigated, it is difficult to access using electromyography [FOR 82].

5.3. Numerical simulation of a non-fractured mandible

The FE method is a valuable tool for understanding the biomechanical behavior of prostheses in their environment. In general, the completion of a numerical simulation can be divided down into several tasks:

– geometric model creation;

– material properties definition;

– meshing model construction;

– boundary condition definition;

– calculation process;

– result treatment.

Rapid developments in calculation software, computer performance and reconstruction methods for anatomical parts such as the mandible have made it possible to create more realistic three-dimensional (3D) models. In the first works of simulating prosthetic dental solutions that drew on the FE method, models that were simplistic in terms of their geometry and applied load were elaborated [NAG 02, PET 05, LAS 08]. In this chapter, we will present complex 3D geometric models with different behaviors and nonlinear boundary conditions.

5.3.1. *Description of the studied mandible*

In order to construct a numerical model of the mandible, tomographic imaging was used for a 3D reconstruction of the mandible's geometry using SOLIDWORK software. In this model, the geometry of the teeth was neglected because the influence of the teeth on the mechanical response is not recognized [KOR 07]. The screws are considered to be cylinders to simplify the geometric modeling. The material of the mandible is modeled as a homogeneous cortical bone with an isotropic behavior. The Young's modulus is 14 GPa and the Poisson ratio is 0.3. For the boundary conditions, Korkmaz [KOR 07] considered the bite force and ignored the muscle forces. The mandible was fixed at its extremities (Figure 5.9).

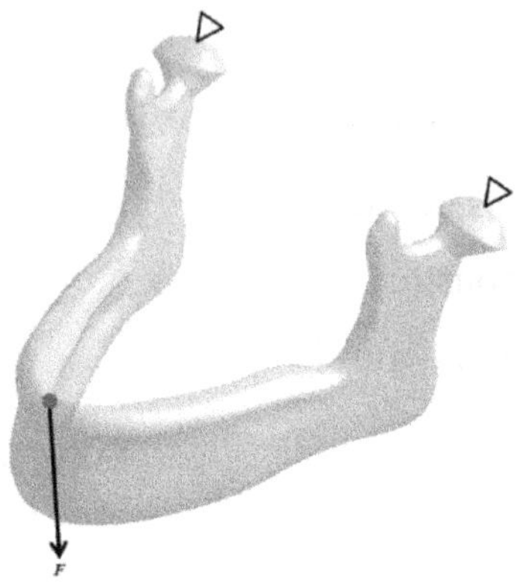

Figure 5.9. *Mandible subject to a bite force and fixed at its extremities*

However, the muscles play a positive role during the bite process [KHA 14, KHA 16b]. Figure 5.10 shows the superficial masseter (SM), deep masseter (DM), posterior temporalis (PT), medial temporalis (MT), anterior temporalis (AT) and medial pterygoid (MP) [RAM 14].

Table 5.1 presents the different values of components of the muscle forces according to Ramos *et al.* [RAM 14].

Muscle forces	F_x [N]	F_y [N]	F_z [N]
Superficial masseter (SM)	18.2	303.3	12.1
Deep masseter (DM)	7.8	128.3	15.6
Anterior temporalis (AT)	–18.4	104.8	–43.8
Medial temporalis (MT)	–6.5	36.3	–53.1
Posterior temporalis (PT)	–3.4	6.8	–37
Medial pterygoid (MP)	187.4	325.1	–76.5

Table 5.1. *Muscle force components*

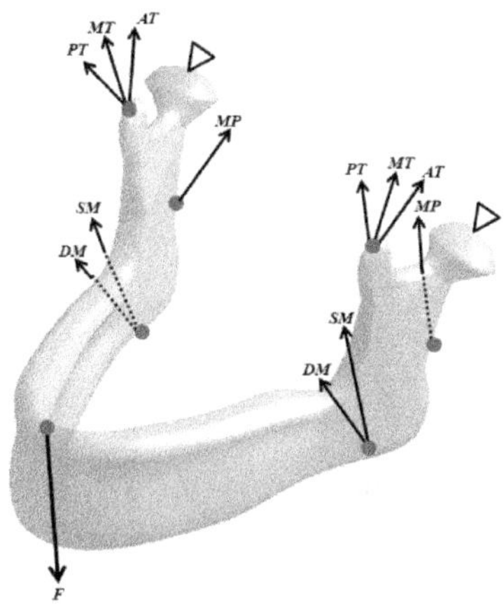

Figure 5.10. *Mandible subject to bite force and muscle forces and fixed at its extremities*

5.3.2. *Numerical results*

During the numerical simulation, the values of the von Mises stresses are considered to be an indication of fracture [GRO 01]. Using two numerical applications in this section, we can then show the importance of muscles during the bite process.

5.3.2.1. *Case of muscle force exclusion*

According to Figure 5.11, the bite force ($F = 208N$) is applied to zone A and the mandible is fixed in zones B and C.

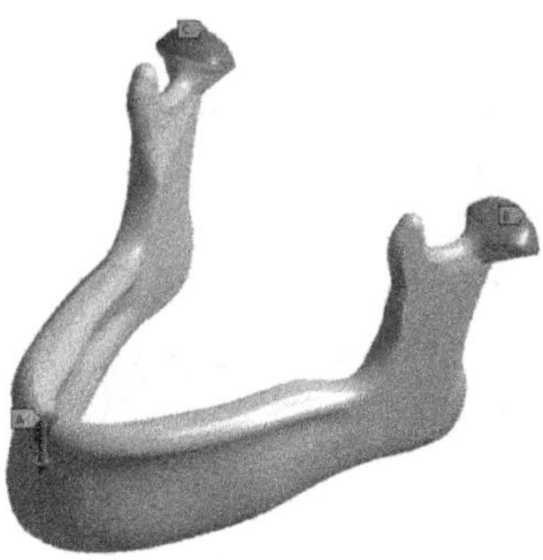

Figure 5.11. *Boundary conditions for the muscle force exclusion case*

Figure 5.12 shows the distribution of von-Mises stresses in the case of excluding muscle forces. The maximum von-Mises stress value is $\sigma_{\max} = 123.74$ MPa [KHA 16b].

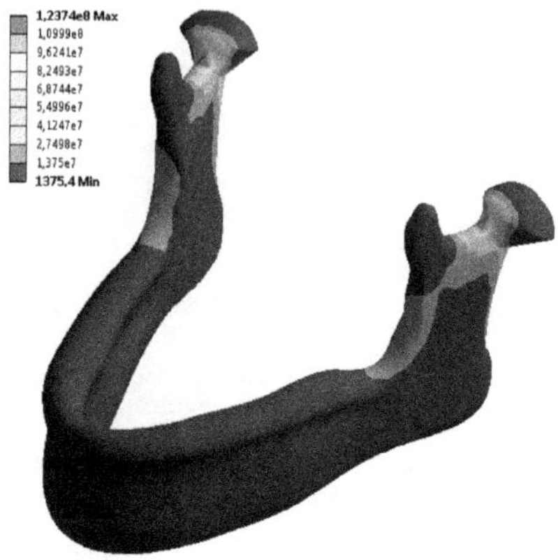

Figure 5.12. *Distribution of the von-Mises stresses in the case of excluding muscle forces. For a color version of the figure, see www.iste.co.uk/kharmanda/biomechanics.zip*

According to the optimized formulation (equation [3.30]) relating the yield stress and Young's modulus and for a given value of Young's modulus ($E = 14\ GPa$), the yield stress in compression is thus $\sigma_C = 160.5$ MPa. Considering the tension/compression ratio is equal to $R_{T/C} = 0.7$ [DOB 04], the yield stress in tension is $\sigma_T = 112.35$ MPa (equation [3.31]). We can note that the value of the maximum von-Mises stress exceeds the yield stress in tension which leads to failure. Due to this, we take the muscle force into account during the numerical study.

5.3.2.2. *Case of muscle force inclusion*

The muscular efforts included in the studied model are the SD, DM, AT, PT and MP. During the bite process, the digastric muscles are not significantly active and thus we can ignore them in this analysis to simplify the study [RAM 14]. According to Figure 5.13, the bite force is applied in region A. The sums of the masseter muscle forces are denoted as M^{Right} and M^{Left} and are applied in regions B and C. The sums of the temporal muscle forces are denoted as T^{Right} and T^{Left} and are applied in regions D and E. Finally, the sums of the pterygoid muscle forces are noted as P^{Right} and P^{Left} and are applied in regions F and G. The mandible is fixed in regions H and I.

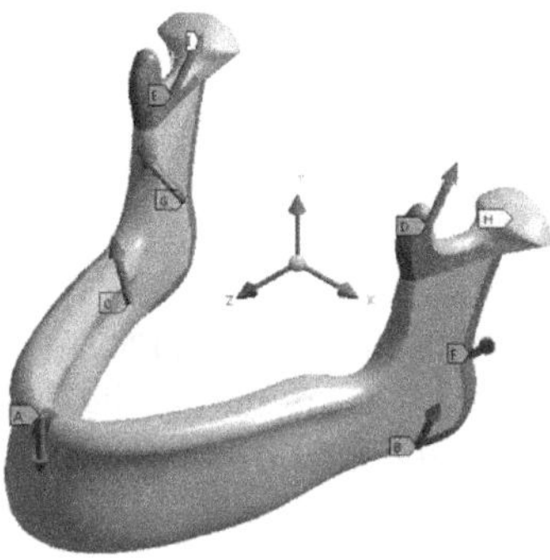

Figure 5.13. *Boundary conditions for the case of muscle force inclusion. For a color version of the figure, see www.iste.co.uk/kharmanda/biomechanics.zip*

Figure 5.14 shows the distribution of the von-Mises stresses in the case of including the muscle forces. The maximum von-Mises stress value is: $\sigma_{\max} = 83.46\ MPa$ [KHA 16b].

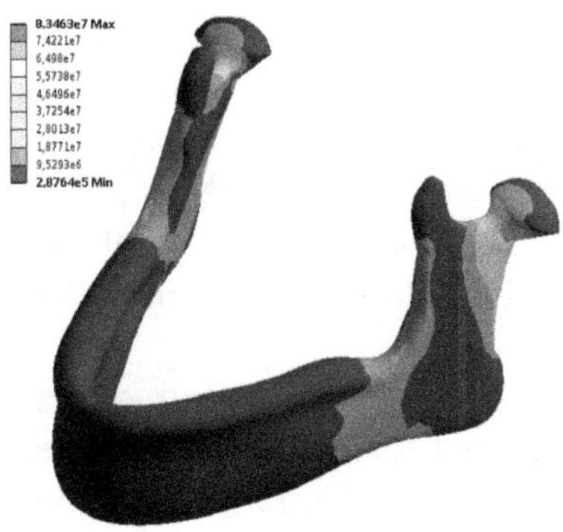

Figure 5.14. *Distribution of von-Mises stresses in the case of including the muscle forces. For a color version of the figure, see www.iste.co.uk/kharmanda/biomechanics.zip*

We can note that the maximum von-Mises stress value does not exceed the yield stress in tension which provides safety. The muscle forces play a positive role during the bite process. Due to this, they should not be ignored during studies. However, during a surgical operation, the muscles can be damaged or cut. The introduction of uncertainty is thus necessary to guarantee the required level of reliability.

5.4. Reliability analysis of the fixation system of the fractured mandible

The fractured mandibles can be treated using a fixation strategy with mini-plates and screws. In general, the behavior of the bone is orthotropic in the case of the mandible [KHA 16c]. Due to this, we present two different cases. The first case is a mandible composed of only one homogeneous cortical bone tissue with an isotropic behavior. The second case is a mandible composed of two bone tissues with an orthotropic behavior.

5.4.1. *Description of a fractured mandible*

Figure 5.15 shows an orthopantomogram photo of a male patient aged 28 [KHA 14, KHA 16b]. The healing period was about 3 months without any specific complication. The fractured part of the mandible is called the symphysis. The reliability analysis in this clinical case was carried out to test the stabilization of the mini-plates used during the healing period following the surgical operation. According to the clinical observations of Cox *et al.* [COX 03], the displacement related to the movement between the two surfaces of the fractures should not exceed a certain value ($150\ \mu m$) to accelerate the healing process.

The reliability analysis of the fractured mandible treated using mini-plates was carried out for two different cases. The two cases can be distinguished by the composition of bone and in the behavior of the material tissue of the bones.

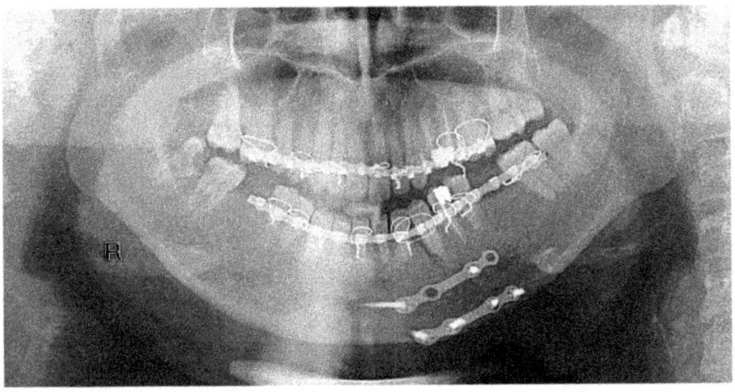

Figure 5.15. *Orthopantomogram photo of a male patient, age 28 [KHA 16b]*

5.4.2. *Fixation strategy using mini-plates*

There are several types of mini-plates: I-plates, L-plates, T-plates and X-plates (Figure 5.16). Korkmaz [KOR 07] carried out a detailed study of several types of mini-plates and provided several recommendations for the selection of type, location and position of the mini-plates.

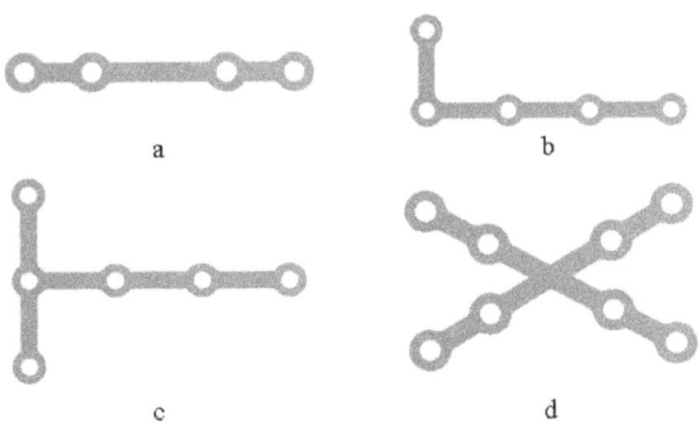

Figure 5.16. *Types of mini-plates*

However, Korkmaz ignored the influence of muscle forces by considering only the bite force. For the clinical case considered here, two I-mini-plates were used with four screws for the lower mini-plate and two screws for the upper mini-plate.

Figure 5.17 shows the dimensions of a mini-plate with a thickness of 2 mm that was used for this surgery.

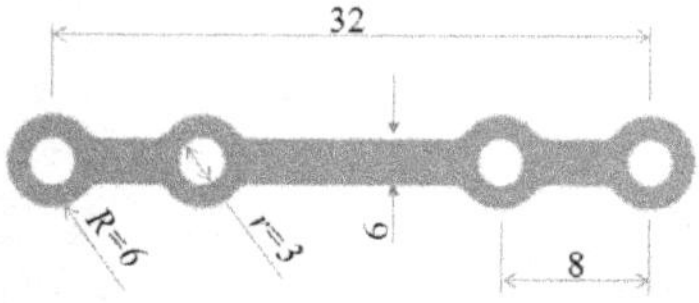

Figure 5.17. *I-Mini-plate*

5.4.3. *Study of a homogeneous and isotropic structure*

5.4.3.1. *Model construction*

The construction of the model includes these steps.

5.4.3.1.1. Geometry and materials

The 3D reconstruction of the geometry of the fractured mandible with mini-plates and screws was completed using SOLIDWORK software, like the study of the non-fractured mandible. The screws are considered to be cylinders to simplify the geometric modeling (Figure 5.18).

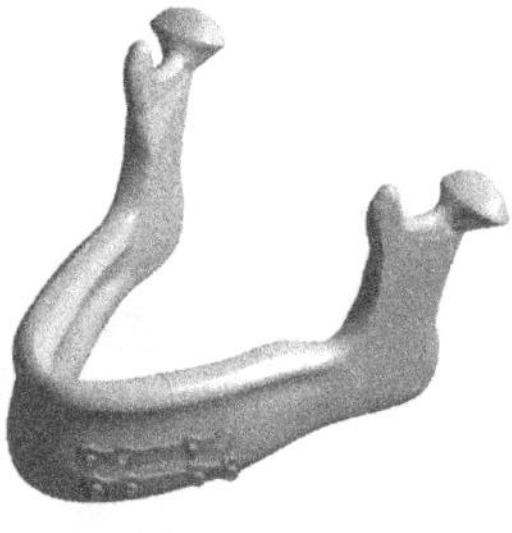

Figure 5.18. *Geometric model of the fractured mandible with mini-plates and screws*

The two parts of the fractured mandible are composed of a homogeneous material of cortical bone with an isotropic behavior where Young's modulus is equal to 14 GPa and the Poisson ratio is equal to 0.3. To calculate the yield stress of the bone, we used an optimized formulation between the elastic limit and the Young's modulus [KHA 15a], which is $\sigma_y^{Bone} = 112.35\,\text{MPa}$.

The mini-plates and the screws are in titanium with a Young's modulus of 110 GPa and a Poisson ratio of 0.34 [KOR 07]. The yield stress of the mini-plates and screws is equal to 860 MPa. The material properties are presented in Table 5.2.

Material	Young's modulus (MPa)	Poisson ratio	Yield stress (MPa)
Titanium	110,000	0.34	860
Cortical bone	14,000	0.3	112.35

Table 5.2. *The material properties of the isotropic study*

5.4.3.1.2. Meshing model

Figure 5.19 shows the meshing model of the studied fractured mandible with the mini-plates and screws using ANSYS. The number of finite elements used is 14,845. The elements used are nonlinear in types: CONTACT and SOLID. The contact between the surfaces of the fracture is considered non-penetrated with an initial distance of 100 μm. The contact between the mini-plates and the screws is completely bonded.

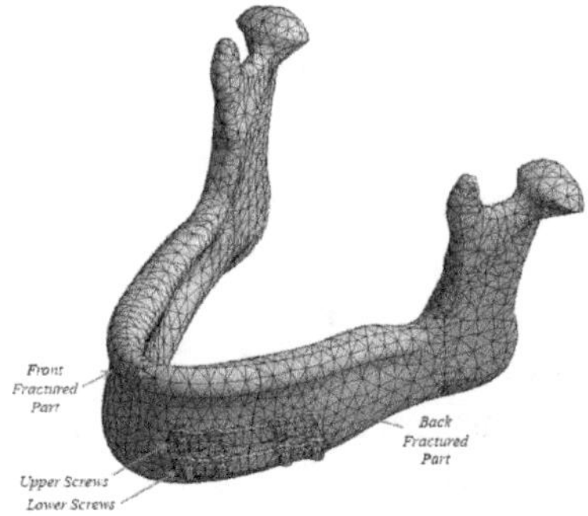

Figure 5.19. *Meshing model of the fractured mandible with mini-plates and screws. For a color version of the figure, see www.iste.co.uk/kharmanda/biomechanics.zip*

5.4.3.1.3. Boundary conditions

The muscle forces included in the studied model are the SM, DM, AT, PT and MP. During the bite process, the digastric muscles are not significantly active and so we can ignore them in this analysis to simplify the study [RAM 14]. According to Figure 5.20, the bite force is applied in region A. The sums of the masseter muscle forces are denoted as M^{Right} and M^{Left} and are applied in regions B and C. The sums of the temporal muscle forces are denoted as T^{Right} and T^{Left} and are applied in regions D and E. Finally, the sums of the pterygoid muscle forces are denoted as P^{Right} and P^{Left} and are applied in regions F and G. The mandible is fixed in regions H and I.

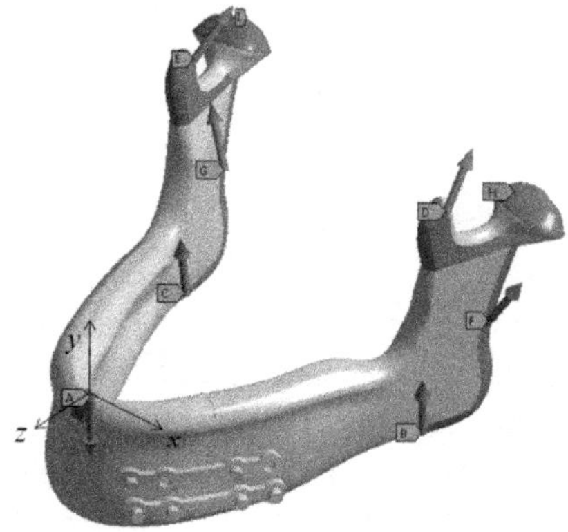

Figure 5.20. *Boundary conditions for the studied homogeneous structure. For a color version of the figure, see www.iste.co.uk/kharmanda/biomechanics.zip*

After the surgical operation, the patient must eat soft foods to have a weak bite force. According to the experimental results of Kumar *et al.* [KUM 13], after the surgery, the maximum value of the postoperation bite force is $F = 44N$. The efforts of the muscles presented in Table 5.1 are measured to the maximum capacity of a non-fractured mandible [MES 11], which corresponds to a bite force of $F = 208N$. Consequently, in the numerical model, the forces in

Table 5.3 are proportionate, so the bite force is $F = 44N$. We consider these values to be the mean values of the problem. In addition, when considering the force uncertainty, it is possible to take into account a confidence interval about several unknown forces where no information is provided.

5.4.3.2. *Developed algorithm*

Three scenarios of failure can occur. Mathematically, we can write the constraints (or limitations) as follows:

$$\begin{aligned} G_1 &= \delta_{\max} - \delta_w \le 0 \\ G_2 &= \sigma_{\max}^{Metal} - \sigma_y^{Metal} \le 0 \\ G_3 &= \sigma_{\max}^{Bone} - \sigma_y^{Bone} \le 0 \end{aligned} \qquad [5.1]$$

where $\delta_{\max}$ is the relative maximum displacement in the area of the fracture δ_w. The parameters $\sigma_{\max}^{Metal}$ and $\sigma_{\max}^{Bone}$, respectively, represent the maximum value of the von-Mises stress effective in the mini-plates and the bone. σ_y^{Metal} and σ_y^{Bone}, respectively, represent the yield stress values in the mini-plates and the bone. The failure is supposed to take place when the constraints $G_1 \le 0$, $G_2 \le 0$ or $G_3 \le 0$ are exceeded. In addition, we can add other constraints to equation [5.1] in order to improve the design. The problem of reliability can be written in this form:

$$\begin{aligned} &\min : d(u_i) = \sqrt{\sum_{i=1}^{n} u_i^2} \\ &s.t.\ : H_1(u_i) = \delta_{\max}(u_i) - \delta_w \le 0 \\ &\quad\ \ : H_2(u_i) = \sigma_{\max}^{Metal}(u_i) - \sigma_y^M \le 0 \\ &\quad\ \ : H_3(u_i) = \sigma_{\max}^{Bone}(u_i) - \sigma_y^M \le 0 \end{aligned} \qquad [5.2]$$

The problem must be solved in a normalized space by considering a transformation of the vector of the random variables y_i into a vector of normalized variables u_i (Figure 5.21).

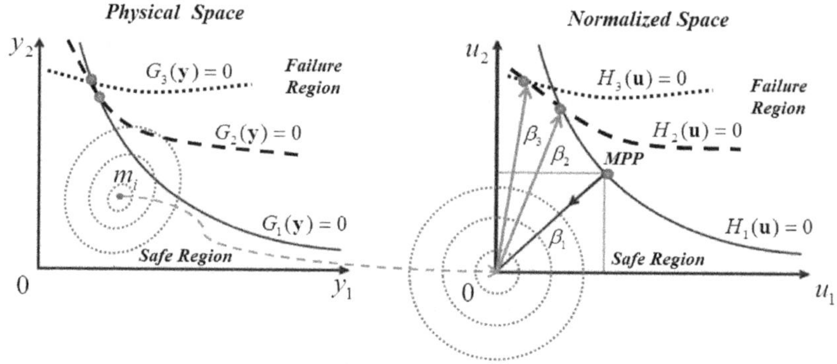

Figure 5.21. *Transformation between the physical space and the normalized space*

Figure 5.22 shows the reliability algorithm developed to find the most probable failure point (MPP). The procedure consists of making two nested loops: the goal of the first loop is to find the force values that correspond to a limit state $G_i(y_i)=0$. Mathematically, the problem posed is to find the first constraint that is exceeded:

$$\max(G_1(y_i),G_2(y_i),G_3(y_i))=0 \qquad [5.3]$$

The second loop is to test the reliability indices β_i for all of the failure points with the goal of finding the MPP that belongs to the limit state $\max(H_1(u_i),H_2(u_i),H_3(u_i))=0$ in the normalized space. The MPP is found when the distance between the origin of the normalized space and the limit state should be minimal.

For the current study, the force values are considered to be mean values. The random variables of these forces are 19 components that are grouped in the vector **y**. In this case, there are 18 muscle forces and 1 bite force. The standard deviations are proportional to the mean values ($\sigma_i = 0.5m_i$).

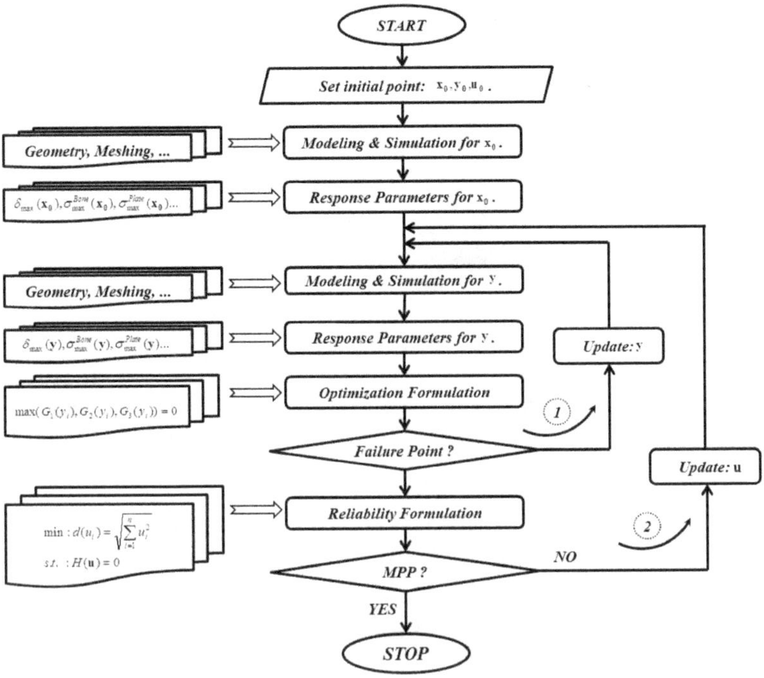

Figure 5.22. *Reliability algorithm*

5.4.3.3. *Numerical results*

After a direct simulation using ANSYS, we found the structural responses corresponding to the mean values where the state of this structure is located in the safe region.

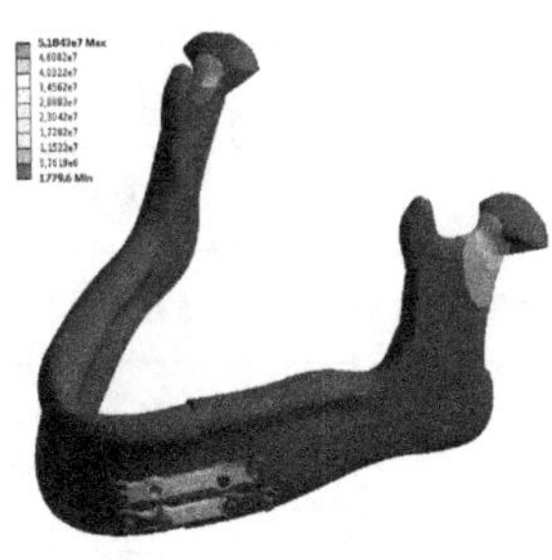

Figure 5.23. *Distribution of the von-Mises stresses. For a color version of the figure, see www.iste.co.uk/kharmanda/biomechanics.zip*

Figure 5.24 shows the distance between the two surfaces of the fractured mandible δ.

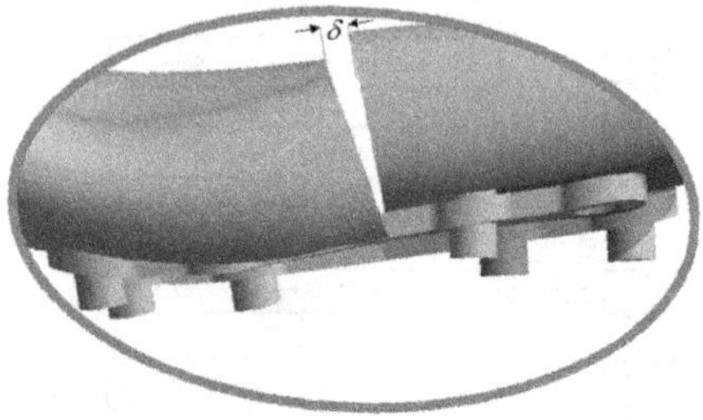

Figure 5.24. *Modeling of relative displacement. For a color version of the figure, see www.iste.co.uk/kharmanda/biomechanics.zip*

First, the algorithm performs the first loop to find several failure points. Then, the reliability index is calculated in the second loop.

Parameters		Means	MPP
F^{Bite} [N]	F_y^{Bite}	–44	–81.66
	M_x^{Right}	5.46	1.47
M^{Right} [N]	M_y^{Right}	90.64	57.48
	M_z^{Right}	5.82	10.64
	M_x^{Left}	–5.46	–5.56
M^{Left} [N]	M_y^{Left}	90.64	41.50
	M_z^{Left}	5.82	9.51
	T_x^{Right}	–5.94	–6.84
T^{Right} [N]	T_y^{Right}	31.06	18.00
	T_z^{Right}	–28.12	–51.10
T^{Left} [N]	T_x^{Left}	5.94	9.45
	T_y^{Left}	31.06	46.42

	T_z^{Left}	–28.12	–20.57
P^{Right} [N]	P_x^{Right}	39.35	45.44
	P_y^{Right}	68.27	75.48
	P_z^{Right}	–16.07	–15.74
P^{Left} [N]	P_x^{Left}	–39.35	–42.66
	P_y^{Left}	68.27	56.88
	P_z^{Left}	–16.07	–19.13

Table 5.3. *The mean values and the MPP*

The results obtained by the algorithm are presented in Tables 5.3 and 5.4. First, considering the loads in Table 5.3, we observe that there is no more symmetry relative to the plane of the origin $x = 0$. The force M_y^{Left} is significantly less than M_y^{Right} at the MPP. We can conclude that the bite force is nearly double its mean. The maximum of the von-Mises stresses is 52 MPa and is found on the lower mini-plate (Figure 5.23).

Limit states	Output parameters	Means	MPP
G_1	$\delta_{\max}$ (μm)	131	149.1
G_2	$\sigma_{\max}^{Lower}$ (MPa)	51.84	65.04
G_2	$\sigma_{\max}^{Upper}$ (MPa)	33.51	48.83
G_3	$\sigma_{\max}^{Left}$ (MPa)	19.07	26.07
G_3	$\sigma_{\max}^{Right}$ (MPa)	17.78	29.34
	β	–	4.17
	P_f	–	1.6×10^{-5}

Table 5.4. *Output parameter values*

At the MPP, the maximum von-Mises stress is equal to 65 MPa, which indicates that there is a large margin relative to the fracture of the mini-plates. The maximum von-Mises stress for the mean values of the forces is equal to 19 MPa. When the MPP is found, the maximum von-Mises stress in the bone is equal to 29 MPa, which also indicates that there is a margin relative to the conditions chosen at the stress level.

However, the problem is related to the relative displacement between the fracture surfaces where its value in the first simulation is $\delta_{\max} = 131\ \mu\text{m}$. For the MPP, we consider that the limit state $\delta_{\max} - \delta_w = 0$ should be satisfied. As a condition for healing, the relative displacement should not exceed a certain value ($\delta_w = 150\ \mu\text{m}$) [COX 03]. Considering the force variability, the reliability index is equal to $\beta = 4.17$, which corresponds to a probability of failure ($P_f \approx 1.6 \times 10^{-5}$).

5.4.4. *Study of a composite and orthotropic structure*

5.4.4.1. *Construction of the model*

The construction of the model follows the following steps.

5.4.4.1.1. Geometry and materials

The 3D reconstruction of the geometry of the fractured mandible in several components with mini-plates and screws was carried out using SOLIDWORK software. The geometry of the teeth was eliminated because the influence of the teeth on the mechanical response is not recognized and the screws are considered to be cylinders to simplify the geometric modeling. The mandible is made up of two parts: the symphysis and the rami. The mandible section is composed of two tissues: cortical bone and trabecular bone (Figure 5.25) [KHA 16c].

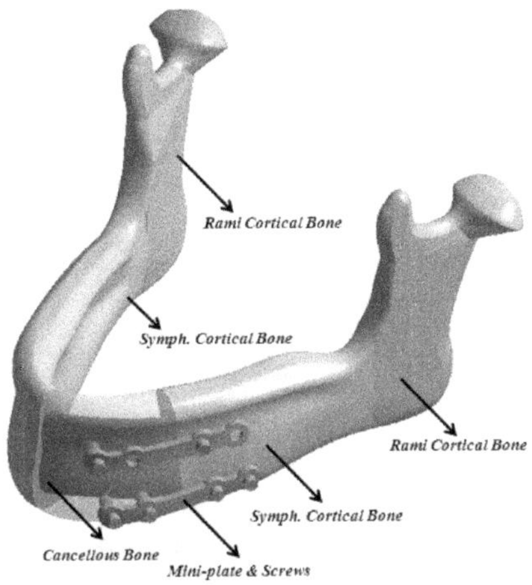

Figure 5.25. *Geometric model of the fractured mandible with mini-plates and screws. For a color version of the figure, see www.iste.co.uk/khamanda/biomechanics.zip*

The cortical bone tissue is considered as an orthotropic material. The material properties are presented in Table 5.5. The equivalent Young's modulus and the equivalent Poisson's ratio are, respectively, calculated according to equations [3.44] and [3.45]. The yield stresses in compression and tension are, respectively, calculated according to equations [3.30] and [3.31].

	Trabecular bone	Symphysis cortical bone	Rami cortical bone
E_x (MPa)	960	22,900	17,000
E_y (MPa)	390	14,200	13,800
E_z (MPa)	320	10,500	10,600
ν_{xy}	0.3	0.19	0.38
ν_{yz}	0.3	0.31	0.23
ν_{xz}	0.3	0.29	0.47
G_{xy} (MPa)	170	6,000	6,200
G_{yz} (MPa)	130	3,700	4,100
G_{xz} (MPa)	90	4,800	5,400

K (MPa)	371.43	10,184.02	13,915.15
G (MPa)	121.5275	4,649.22	5,081.49
E (MPa)	328.73	12,105.51	13,590.20
ν	0.35	0.30	0.34
σ_C (MPa)	4.72	140	156
σ_T (MPa)	3.30–4.72	70–98	78–109

Table 5.5. *The material properties in the orthotropic study*

The mini-plates and screws are made of titanium with a Young's modulus of 110 GPa and a Poisson ratio of 0.34 [KOR 07]. The yield stress of the mini-plates and screws is 860 MPa.

5.4.4.1.2. Meshing model

Figure 5.26 shows a meshing model of the fractured mandible studied with the mini-plates and screws. The number of finite elements used is 35,528 (nonlinear elements of the types: CONTACT and SOLID). The contact between the surfaces of the fracture is considered non-penetrated with a distance of $100\,\mu m$. The contact between the mini-plates and screws is completely bonded. The contact between the cortical bone and trabecular (or cancellous) bone is also completely bonded.

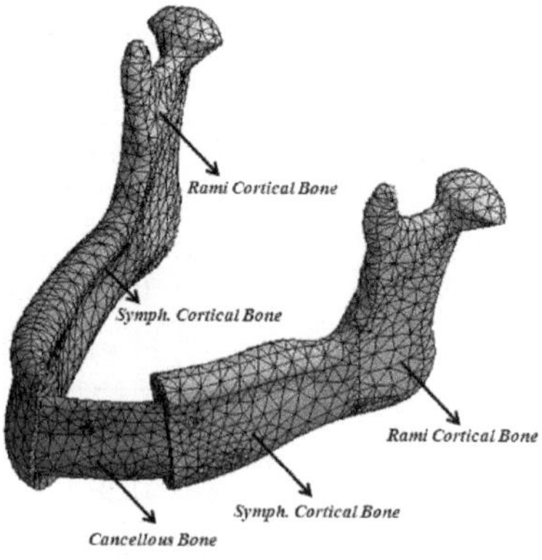

Figure 5.26. *Meshing model of the studied fractured mandible with mini-plates and screws. For a color version of the figure, see www.iste.co.uk/kharmanda/biomechanics.zip*

5.4.4.1.3. Boundary conditions

The bite force is applied in region A. The sums of the masseter muscle forces are denoted as M^{Right} and M^{Left} and are applied in regions B and C. The sums of the temporal muscle forces are denoted as T^{Right} and T^{Left} and are applied in regions D and E. Finally, the sums of the pterygoid muscle forces are denoted as P^{Right} and P^{Left} and are applied in regions F and G. The mandible is fixed in regions H and I.

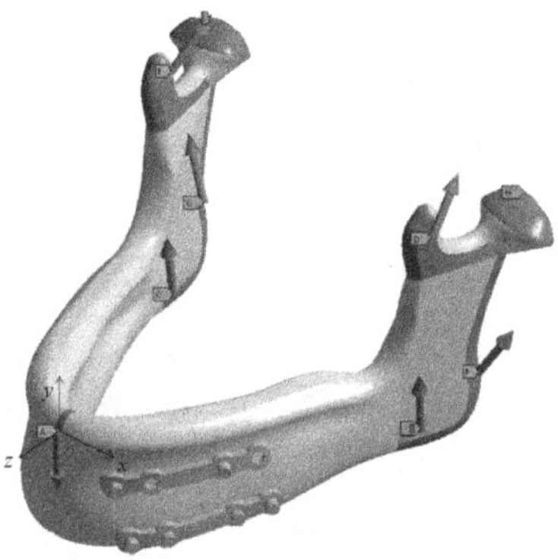

Figure 5.27. *Boundary conditions of the studied fractured mandible with composite bone tissues. For a color version of the figure, see www.iste.co.uk/kharmanda/biomechanics.zip*

We consider these values to be the mean values of our probabilistic problem. When considering the force uncertainty, we have a confidence interval taking into account the different unknown forces.

5.4.4.2. *Developed algorithm*

There are five scenarios of failure that may occur. Mathematically, we can write the constraints as follows:

$$
\begin{aligned}
G_1 &= \delta_{\max} - \delta_w \le 0 \\
G_2 &= \sigma_{\max}^{Metal} - \sigma_y^{Metal} \le 0 \\
G_3 &= \sigma_{\max}^{SymCorBone} - \sigma_y^{SymCorBone} \le 0 \\
G_4 &= \sigma_{\max}^{RamiCorBone} - \sigma_y^{RamiCorBone} \le 0 \\
G_5 &= \sigma_{\max}^{TrabBone} - \sigma_y^{TrabBone} \le 0
\end{aligned}
\qquad [5.4]
$$

where $\delta_{\max}$ is the maximum relative displacement in the fracture zone. The parameters $\sigma_{\max}^{Metal}$ and σ_{y}^{Metal} represent the maximum value of the von-Mises stress and the yield stress of the mini-plates and the screws, respectively. The parameters $\sigma_{\max}^{SymCorBone}$ and $\sigma_{\max}^{RamCorBone}$ represent the maximum value of the von-Mises stress effective in the components of the bones in the symphysis and the rami, respectively. In addition, the parameters $\sigma_{y}^{SymCorBone}$ and $\sigma_{y}^{RamCorBone}$ represent the yield stress in tension in the components of the bones in the symphysis and the rami, respectively. The parameters $\sigma_{\max}^{TrabBone}$ and $\sigma_{y}^{TrabBone}$ represent the maximum von-Mises stress value and the yield stress in tension of the trabecular bone.

The failure is supposed to occur when the constraints $G_1 \leq 0$, $G_2 \leq 0$, $G_3 \leq 0$, $G_4 \leq 0$ or $G_5 \leq 0$ are violated. In addition, we can add other constraints to equation [5.4] to improve its design. The reliability problem can be written as follows:

$$
\begin{aligned}
&\min : d(u_i) = \sqrt{\sum_{i=1}^{n} u_i^2} \\
&s.t. \;: H_1(u_i) = \delta_{\max}(u_i) - \delta_w \leq 0 \\
&\quad\;\; : H_2(u_i) = \sigma_{\max}^{Metal}(u_i) - \sigma_{y}^{Metal} \leq 0 \\
&\quad\;\; : H_3(u_i) = \sigma_{\max}^{SymCorBone}(u_i) - \sigma_{y}^{SymCorBone} \leq 0 \\
&\quad\;\; : H_4(u_i) = \sigma_{\max}^{RamiCorBone}(u_i) - \sigma_{y}^{RamiCorBone} \leq 0 \\
&\quad\;\; : H_5(u_i) = \sigma_{\max}^{TrabBone}(u_i) - \sigma_{y}^{TrabBone} \leq 0
\end{aligned}
\qquad [5.5]
$$

The problem must be solved in the normalized space by considering a transformation of the vector of the random variables y_i to a vector of normalized variables u_i (Figure 5.28).

Figure 5.29 shows the reliability algorithm developed to find the MPP by taking into account the violation of one of the constraints $G_i(y_i) = 0$. Mathematically, the problem posed is to find the first constraint to be violated:

$$\max(G_1(y_i), G_2(y_i), G_3(y_i), G_4(y_i), G_5(y_i)) = 0 \qquad [5.6]$$

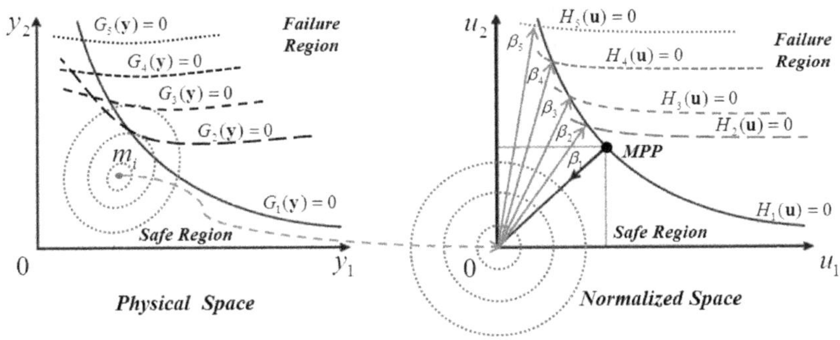

Figure 5.28. *Transformation between the physical space and normalized space*

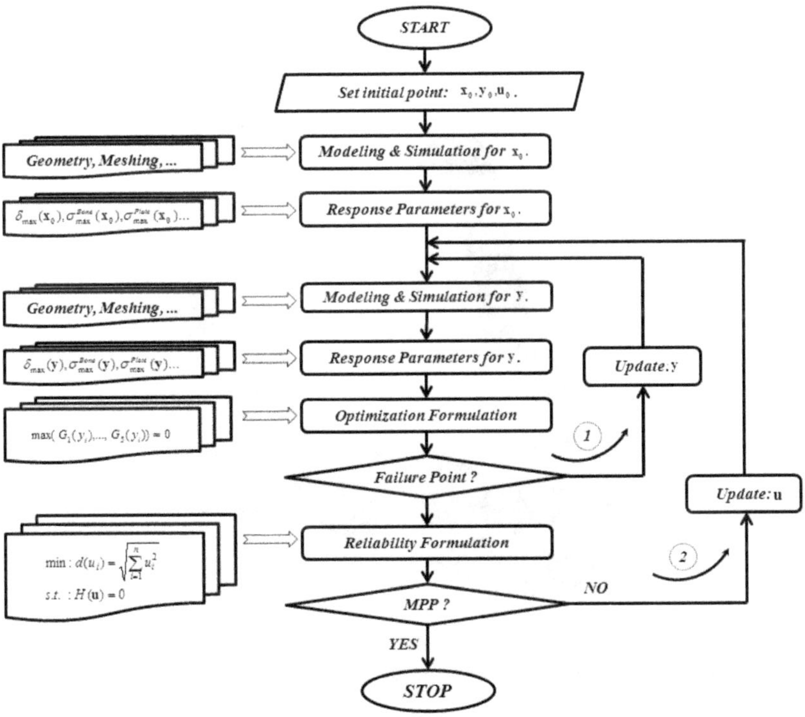

Figure 5.29. *Reliability algorithm*

The second loop is performed in order to test the reliability indices β_i for all of the failure points with the object of finding the MPP that belongs to the limit state $\max(H_1(u_i), H_2(u_i), H_3(u_i), H_4(u_i), H_5(u_i)) = 0$ in the normalized space. The MPP is found when the distance between the origin of the normalized space and the limit state is minimal.

For the current study, the force values are considered to be mean values. The random variables of these forces are 19 components that are grouped in the vector **y**. In this case, there are 18 muscle forces and 1 bite force. The standard deviations are assumed to be proportionate to the mean values ($\sigma_i = 0.25\, m_i$).

5.4.4.3. *Numerical results*

After a direct simulation using ANSYS, we found the structural responses corresponding to the mean values where the current state of this structure is in safety region.

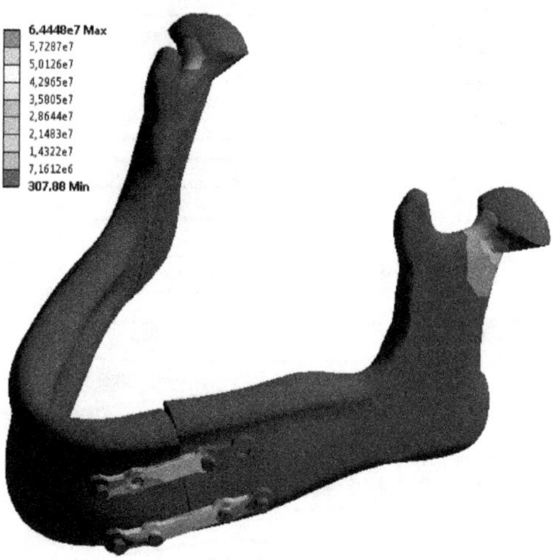

Figure 5.30. *Distribution of the von-Mises stresses. For a color version of the figure, see www.iste.co.uk/kharmanda/biomechanics*

First, the algorithm performs the first loop to find several failure points. Then, the reliability index is calculated in the second loop.

Parameters		Means	MPP
F^{Bite} [N]	F_y^{Bite}	–44	–56.08
M^{Right} [N]	M_x^{Right}	5.46	6.15
	M_y^{Right}	90.64	88.96
	M_z^{Right}	5.82	6.14
M^{Left} [N]	M_x^{Left}	–5.46	–6.20
	M_y^{Left}	90.64	68.70
	M_z^{Left}	5.82	5.93
T^{Right} [N]	T_x^{Right}	–5.94	–6.29
	T_y^{Right}	31.06	31.35
	T_z^{Right}	–28.12	–30.67
T^{Left} [N]	T_x^{Left}	5.94	5.64
	T_y^{Left}	31.06	28.41
	T_z^{Left}	–28.12	–25.60
P^{Right} [N]	P_x^{Right}	39.35	58.95
	P_y^{Right}	68.27	63.61
	P_z^{Right}	–16.07	–14.19
P^{Left} [N]	P_x^{Left}	–39.35	–56.08
	P_y^{Left}	68.27	65.66
	P_z^{Left}	–16.07	–16.48

Table 5.6. *The mean value and the MPP*

The results obtained using the algorithm are shown in Tables 5.6 and 5.7. First, when considering the loads in Table 5.6, we observe that there is no more symmetry relative to the plane of the origin $x = 0$. The force M_y^{Left} is significantly less than M_y^{Right} at the MPP. We can conclude that the bite force is nearly 30% in relation to its mean. The maximum von-Mises stress is 64 MPa on the lower mini-plate (Figure 5.30).

Output parameters	Means	MPP
$\sigma_{\max}^{Metal\_Upper}$ (MPa)	35.94	53.04
$\sigma_{\max}^{Metal\_Lower}$ (MPa)	64.45	94.47
$\sigma_{\max}^{RamiCor\_Right}$ (MPa)	22.62	29.53
$\sigma_{\max}^{RamiCor\_Left}$ (MPa)	18.36	21.81
$\sigma_{\max}^{SymCor\_RightBack}$ (MPa)	12.64	17.53
$\sigma_{\max}^{SymCor\_RightFront}$ (MPa)	6.28	15.96
$\sigma_{\max}^{SymCor\_Left}$ (MPa)	4.66	6.04
$\sigma_{\max}^{Trab\_RightBack}$ (MPa)	0.17	0.24
$\sigma_{\max}^{Trab\_RightFront}$ (MPa)	0.24	0.40
$\sigma_{\max}^{Trab\_Left}$ (MPa)	0.10	0.13
$\delta_{\max}^{Cor}$ (μm)	132	149
$\delta_{\max}^{Trab}$ (μm)	116	125
β	–	3.41
P_f	–	3.2×10^{-4}

Table 5.7. *The output parameters of the reliability analysis*

At the MPP, the maximum von-Mises stress is 94 MPa, which indicates that there is a large margin relative to the fracture of the mini-plates. The maximum von-Mises stress for the mean values of the forces is 23 MPa. When the MPP is found, the maximum von-Mises stress in the bone is 30 MPa, which also indicates that there is a margin relative to the conditions chosen at the stress level.

However, the problem is related to the relative displacement between the fracture surfaces where its value in the first simulation $\delta_{\max} = 132\ \mu\text{m}$. For the MPP, we consider that the limit state $\delta_{\max} - \delta_w = 0$ must be satisfied. As a condition of healing, the relative displacement must not exceed a certain value ($\delta_w = 150\ \mu\text{m}$) [COX 03]. By considering the force variability, the reliability index is equal to $\beta = 3.41$, which corresponds to a probability of failure ($P_f \approx 1.6 \times 10^{-5}$).

5.4.5. *Result discussion*

The primary goal of managing a fracture is to reduce the healing time of a fractured bone in such a way that the shape and load make it possible to restore the fractured bone. A common strategy uses fixation through surgery where mini-plates are used to stabilize the fractured parts. During the surgery, the muscles can be damaged. In this case, it is difficult for the muscles to function at their maximum capacity. Therefore, uncertainty must be incorporated into the muscle forces. It is also important to integrate uncertainty into the material properties and the geometry of the mandible. We can conclude that determinist methods cannot be compatible with managing the fracture.

Several failure scenarios can occur during the healing process. The maximum stress of the components of the bone and metal should not exceed the yield stress. In addition, to simplify the restoration of the bone of the fractured interface, the relative displacement between the surfaces of the fracture should not exceed an admissible value ($\delta_{\max} \le 150\mu m$). The reliability formula is general and can be easily extended to other constraints. Other failure scenarios can be described by the detachment of the screws which leads to a failure during the healing process. The algorithm is able to find the different failure scenarios. This advantage can help to improve the fixation system. In the studied cases, we found that the relative displacement is the most active stress. Another observation about the reliability analysis is that the muscle forces are not symmetrical. This observation is logical because muscle activity is never symmetrical, which requires the introduction of uncertainties during the design process. It should be noted that several simplifications were taken into account: geometric simplification, static load and boundary conditions.

5.5. Conclusion

The incorporation of reliability is a new scientific area to better understand the different uncertainties in biomechanics. In this chapter, the principle of integration was carried out on orthodontic prostheses, in particular on the treatment of a fractured mandible. First, an introduction to the anatomy of the temporomandibular joint was presented to demonstrate the function of the different muscles. Then, a numerical simulation of a non-fractured mandible was presented by considering two cases of boundary conditions. We found that the muscles play a positive role in the stabilization of the fixation system. Finally, two numerical applications were carried out to evaluate the reliability levels of the fixation systems used when considering two behavior cases of the bone materials.

Appendices

Appendix 1

Matrix Calculation

A1.1. Linear equations system

Linear equations can be written in the following form:

$$a_{11}x_1 + a_{12}x_2 + \ldots + a_{1n}x_n = b_1$$

$$a_{21}x_1 + a_{22}x_2 + \ldots + a_{2n}x_n = b_2$$

$$a_{n1}x_1 + a_{n2}x_2 + \ldots + a_{nn}x_n = b_n$$

The unknown values are: $x_1, x_2, \ldots, x_n$.

It can be also written in the following matrix form:

$$\mathbf{AX} = \mathbf{b}$$

The square matrix of the coefficients can be written as follows:

$$A = \left[a_{ij}\right] = \begin{bmatrix} a_{11} & a_{12} & \cdots & a_{1n} \\ a_{21} & a_{22} & \cdots & a_{2n} \\ \cdots & \cdots & \cdots & \cdots \\ a_{n1} & a_{n2} & \cdots & a_{nm\ldots} \end{bmatrix},$$

The vector of the unknown values can be written as follows:

$$\mathbf{X} = \{x_i\} = \begin{Bmatrix} x_1 \\ x_2 \\ \vdots \\ x_n \end{Bmatrix},$$

Moreover, the constants vector can be written in the following form:

$$\mathbf{b} = \{b_i\} = \begin{Bmatrix} b_1 \\ b_2 \\ \vdots \\ b_n \end{Bmatrix}$$

A1.2. Addition and subtraction of matrices

The addition and subtraction of matrices having the same dimensions can be carried out using the following form:

$$\mathbf{C} = \mathbf{A} + \mathbf{B} \quad \text{with} \quad c_{ij} = a_{ij} + b_{ij}$$

$$\mathbf{D} = \mathbf{A} - \mathbf{B} \quad \text{with} \quad d_{ij} = a_{ij} - b_{ij}$$

A1.3. Scalar multiplication

The multiplication of a matrix by a scalar is carried out using the following form:

$$\lambda\mathbf{A} = \left[\lambda a_{ij}\right]$$

A1.4. Product of matrices

The product of two matrices is carried out as follows:

$$\mathbf{C} = \mathbf{AB} \quad \text{with} \quad c_{ij} = \sum_{k=1}^{m} a_{ik} b_{kj}, \; i = 1,2,\ldots,l; \;\; j = 1,2,\ldots,n$$

The product of two matrices is not commutative:

$$\mathbf{AB} \neq \mathbf{BA}$$

The product of two matrices is associative:

$$(\mathbf{AB})\mathbf{C} = \mathbf{A}(\mathbf{BC})$$

A1.5. Transpose of a matrix

The transposed matrix of a matrix $\mathbf{A} = \left[a_{ij}\right]$ can be written using the following form:

$$\mathbf{A}^T = \left[a_{ji}\right]$$

The transpose of the product of two matrices is equal to the product of the transpose of the second matrix by the transpose of the first matrix, which is expressed by the following form:

$$(\mathbf{AB})^T = \mathbf{B}^T\mathbf{A}^T$$

A1.6. Symmetric matrix

The transpose of a symmetric matrix is equal to the same matrix by considering the following equation:

$$\mathbf{A} = \mathbf{A}^T, \quad a_{ij} = a_{ji}$$

A1.7. Unit matrix

The product of a unit matrix by another matrix is equal to the same matrix and we have:

$$\mathbf{AI} = \mathbf{A} \ , \mathbf{IX} = \mathbf{X}$$

where the unit matrix can be written as follows:

$$\mathbf{I} = \begin{bmatrix} 1 & 0 & \dots & 0 \\ 0 & 1 & \dots & 0 \\ \dots & \dots & \dots & \dots \\ 0 & 0 & \dots & 1 \end{bmatrix}$$

A1.8. Determinant of a matrix

The determinant of a matrix $|\mathbf{A}|$ (or $\det \mathbf{A}$) is a scalar value. For example, for a two-dimensional square matrix, we have:

$$\det \begin{bmatrix} a & b \\ c & d \end{bmatrix} = ad - bc$$

and for a three-dimensional square matrix, we have:

$$\det \begin{bmatrix} a_{11} & a_{12} & a_{13} \\ a_{21} & a_{22} & a_{23} \\ a_{31} & a_{32} & a_{33} \end{bmatrix} =$$
$$= a_{11}a_{22}a_{33} + a_{12}a_{23}a_{31} + a_{21}a_{32}a_{13} - a_{13}a_{22}a_{31} - a_{12}a_{21}a_{33} - a_{23}a_{32}a_{11}$$

A1.9. Singular matrix

When a determinant of a matrix is nil ($\det \mathbf{A} = 0$), the matrix is considered singular. In this case, we have several solutions.

A1.10. Inverse matrix

For a square matrix $\mathbf{A}$, if the determinant is not nil and the matrix itself is not singular, the inverse matrix $\mathbf{A}^{-1}$ verifies the following condition:

$$\mathbf{A}\mathbf{A}^{-1} = \mathbf{A}^{-1}\mathbf{A} = \mathbf{I}$$

It can be determined using the assistant matrix:

$$c_{ij} = (-1)^{i+j} M_{ij}$$

wherein M_{ij} is the determinant of the submatrices during elimination of line number i and the column number j in the principal matrix $\mathbf{A}$. Then, the inverse matrix is calculated as follows:

$$\mathbf{A}^{-1} = \frac{1}{\det \mathbf{A}} \mathbf{C}^T$$

which verifies the following condition:

$$(\mathbf{AB})^{-1} = \mathbf{B}^{-1}\mathbf{A}^{-1}$$

ILLUSTRATIVE EXAMPLE A1.1.–

In the case of a two-dimensional square matrix, we have:

$$\begin{bmatrix} a & b \\ c & d \end{bmatrix}^{-1} = \frac{1}{(ad-bc)} \begin{bmatrix} d & -b \\ -c & a \end{bmatrix}$$

To verify, we use the unit matrix:

$$\mathbf{AA}^{-1} = \mathbf{A}^{-1}\mathbf{A} = \mathbf{I}$$

$$\begin{bmatrix} a & b \\ c & d \end{bmatrix}^{-1} \begin{bmatrix} a & b \\ c & d \end{bmatrix} = \frac{1}{(ad-bc)} \begin{bmatrix} d & -b \\ -c & a \end{bmatrix} \begin{bmatrix} a & b \\ c & d \end{bmatrix} = \begin{bmatrix} 1 & 0 \\ 0 & 1 \end{bmatrix}$$

In the case of a three-dimensional square matrix, we have:

$$\begin{bmatrix} 1 & -1 & 0 \\ -1 & 2 & -1 \\ 0 & -1 & 2 \end{bmatrix}^{-1} = \frac{1}{(4-2-1)} \begin{bmatrix} 3 & 2 & 1 \\ 2 & 2 & 1 \\ 1 & 1 & 1 \end{bmatrix}^T = \begin{bmatrix} 3 & 2 & 1 \\ 2 & 2 & 1 \\ 1 & 1 & 1 \end{bmatrix}$$

To verify, we use the unit matrix:

$$\mathbf{AA}^{-1} = \mathbf{A}^{-1}\mathbf{A} = \mathbf{I}$$

$$\begin{bmatrix} 1 & -1 & 0 \\ -1 & 2 & -1 \\ 0 & -1 & 2 \end{bmatrix}\begin{bmatrix} 3 & 2 & 1 \\ 2 & 2 & 1 \\ 1 & 1 & 1 \end{bmatrix} = \begin{bmatrix} 1 & 0 & 0 \\ 0 & 1 & 0 \\ 0 & 0 & 1 \end{bmatrix}$$

The calculation of the inverse matrix can be used in several numerical methods especially optimization. For example, for second-order optimization methods, we calculate the inverse matrix of the second derivative. Moreover, for the numerical methods used to solve linear equations system, we calculate the inverse matrix of the matrix of the coefficients:

$$\mathbf{X} = \mathbf{A}^{-1}\mathbf{b}$$

Appendix 2

ANSYS Code for the Disk Implant

```
!****************************************************
!*** Start ANSYS Code
!*** Disk Implant
!*** Input Parameters
*SET,d1,50
*SET,d2,46
!***************
!*** Preprocessor
!***************
/PREP7
!**** Geometry Creation
K,1,0,-5,0,
K,2,10,-5,0,
K,3,20,-5,0,
K,4,d2/2,-3,0,
K,5,d1/2,0,0,
K,6,d2/2,3,0,
K,7,20,5,0,
K,8,10,5,0,
K,9,0,5,0,
K,10,-10,5,0,
K,11,-20,5,0,
K,12,-d2/2,3,0,
K,13,-d1/2,0,0,
K,14,-d2/2,-3,0,
K,15,-20,-5,0,
K,16,-10,-5,0,
K,16,-10,-5,0,
```

```
LSTR,      1,      2
LSTR,      2,      3
LSTR,      7,      8
LSTR,      8,      9
LSTR,      9,     10
LSTR,     10,     11
LSTR,     15,     16
LSTR,     16,      1
FLST,3,5,3
FITEM,3,3
FITEM,3,4
FITEM,3,5
FITEM,3,6
FITEM,3,7
BSPLIN, ,P51X
FLST,3,5,3
FITEM,3,11
FITEM,3,12
FITEM,3,13
FITEM,3,14
FITEM,3,15
BSPLIN, ,P51X
LSTR,      1,      9
LSTR,      2,      8
LSTR,     16,     10
FLST,2,4,4
FITEM,2,1
FITEM,2,12
FITEM,2,4
FITEM,2,11
AL,P51X
FLST,2,4,4
FITEM,2,12
FITEM,2,2
FITEM,2,9
FITEM,2,3
AL,P51X
FLST,2,4,4
FITEM,2,8
FITEM,2,11
FITEM,2,5
FITEM,2,13
AL,P51X
```

```
FLST,2,4,4
FITEM,2,7
FITEM,2,13
FITEM,2,6
FITEM,2,10
AL,P51X
!*** Element Type Definition
ET,1,PLANE42
!*** Material Properties
MPTEMP,,,,,,,,
MPTEMP,1,0
MPDATA,EX,1,,100000
MPDATA,PRXY,1,,0.2
MPTEMP,,,,,,,,
MPTEMP,1,0
MPDE,EX,1
MPDE,PRXY,1
MPDATA,EX,1,,100000
MPDATA,PRXY,1,,0.2
!*** Meshing Creation
MSHAPE,0,2D
MSHKEY,0
FLST,5,4,5,ORDE,2
FITEM,5,1
FITEM,5,-4
CM,_Y,AREA
ASEL, , , ,P51X
CM,_Y1,AREA
CHKMSH,'AREA'
CMSEL,S,_Y
AMESH,_Y1
CMDELE,_Y
CMDELE,_Y1
CMDELE,_Y2
FINISH
!*****************
!*** Solution Process
!*****************
/SOLU
!*** Analysis Type Definition
```

```
ANTYPE,0
!*** Boundary Condition Definition
FLST,2,4,4,ORDE,4
FITEM,2,1
FITEM,2,-2
FITEM,2,7
FITEM,2,-8
/GO
DL,P51X, ,ALL,
FLST,2,4,4,ORDE,2
FITEM,2,3
FITEM,2,-6
/GO
SFL,P51X,PRES,10,
/STATUS,SOLU
SOLVE
!***************
!*** Postprocessor
!***************
FINISH
/POST1
PLNSOL, S,EQV, 0,1.0
!*** Volume Evaluation
AVPRIN,0, ,
ETABLE,EVOL,VOLU,
SSUM
*GET,VOLUME,SSUM, ,ITEM,EVOL
!*** Maximum Stress Evaluation
AVPRIN,0, ,
ETABLE,SMAX,S,EQV
NSORT,S,EQV,0,0, ,0
*GET,SMAX,SORT,,MAX
!***End ANSYS Code
!**************************************************
```

Appendix 3

ANSYS Code for the Stem Implant

```
!*************************************************
!*** Start ANSYS Code
!*** Stem Implant
!*** Input Parameters
!************
*SET,M_EM,110000
*SET,M_ECOR,17000
*SET,M_ECAN,761
*SET,M_NUM,0.3
*SET,M_NUCOR,0.3
*SET,M_NUCAN,0.3
!************
*SET,FB1X,0
*SET,FB1Y,0
*SET,FB2X,12.8335
*SET,FB2Y,8.6563
*SET,FB3X,0
*SET,FB3Y,0
!************
*SET,FM1X,2.6843
*SET,FM1Y,3.8336
*SET,EM,100000
*SET,ECOR,14000
*SET,ECAN,192
*SET,NUM,0.3
*SET,NUCOR,0.3
*SET,NUCAN,0.3
!************
```

```
*SET,LL,2
*SET,LLR,2
*SET,LLH,2
!***************
!*** Preprocessor
!***************
/PREP7
!**** Formulation
*SET,S_EM,0.1*M_EM
*SET,S_ECOR,0.1*M_ECOR
*SET,S_ECAN,0.1*M_ECAN
*SET,S_NUM,0.1*M_NUM
*SET,S_NUCOR,0.1*M_NUCOR
*SET,S_NUCAN,0.1*M_NUCOR
!************
*SET,U_EM,(EM-M_EM)/S_EM
*SET,U_ECOR,(ECOR-M_ECOR)/S_ECOR
*SET,U_ECAN,(ECAN-M_ECAN)/S_ECAN
*SET,U_NUM,(NUM-M_NUM)/S_NUM
*SET,U_NUCOR,(NUCOR-M_NUCOR)/S_NUCOR
*SET,U_NUCAN,(NUCAN-M_NUCAN)/S_NUCAN
!************
DIS01=U_EM**2+U_ECOR**2+U_ECAN**2
DIS02=U_NUM**2+U_NUCOR**2+U_NUCAN**2
!************
BETA=SQRT(DIS01+DIS02)
!************
YS_COR=129.5*(10**((0.94)*LOG10(ECOR/11142)))
!************
YS_CAN=129.5*(10**((0.94)*LOG10(ECAN/11142)))
!************
ST_COR=SQRT(2*ECOR*0.00000503)
ST_CAN=SQRT(2*ECAN*0.00000503)
!************
*SET,L,LL-2
*SET,LR,LLR-2
*SET,LH,LLH-2
!**** Geometry Creation
K,1,   -2.74022868,       0.00000+LH,      0.00000,
K,2, -4.24022868,      1.50000+LH,      0.00000,
K,3, -4.24022868,      1.90365954+LH,   0.00000,
K,4, -4.24022868,      48.99389158,   0.00000,
K,5, -6.2633839,      59.83419428,   0.00000,
```

```
K,6, -11.93473356,     66.99088707,  0.00000,
K,7, -14.76316068,     69.81931419,  0.00000,
K,8, -17.55361021-L,    67.02886466-L,  0.00000,
K,9, -19.67493056-L,    69.15018501-L,  0.00000,
K,10, -19.67493056-L,    70.56439857-L,  0.00000,
K,11, -15.41226443,     74.82706469,  0.00000,
K,12, -17.05012644,     79.42409881,  0.00000,
K,13, -23.73262659,     89.93009367,  0.00000,
K,14, -38.93542238,     105.13288946, 0.00000,
K,15, -29.74303423,     114.32527762, 0.00000,
K,16, -14.54023843,     99.12248182,  0.00000,
K,17, -6.23642786,     94.13549109,  0.00000,
K,18, 2.04497531,     92.28430443,  0.00000,
K,19, 6.8287523+LR,     97.06808142+LR,  0.00000,
K,20, 8.24296586+LR,     97.06808142+LR,  0.00000,
K,21, 10.3642862+LR,     94.94676108+LR,  0.00000,
K,22, 7.53585908,     92.11833396,  0.00000,
K,23, 10.3642862,     89.28990683,  0.00000,
K,24, 11.54736632,     87.73931554,  0.00000,
K,25, 13.04844415,     82.95202845,  0.00000,
K,26, 13.16305675,     75.34672931,  0.00000,
K,27, 7.9238176,     48.99389158,  0.00000,
K,28, 4.27599184,     1.90365954+LH,  0.00000,
K,29, 4.23574826,     1.38415018+LH,  0.00000,
K,30, 2.74022868,     0.00000+LH,    0.00000,
K,31, -2.74022868,     1.50000+LH,    0.00000,
K,32, 2.74022868,     1.50000+LH,    0.00000,
K,33, 7.53585908+LR,     96.36097464+LR,  0.00000,
K,34, -18.96782378-L,    69.85729179-L,  0.00000,
!***********************************************
KBETW,15,14,0,RATI,0.5,
K, 101 ,14.3+4,98.8+4,0
K, 102 ,18.2+3,103.35+3,0
K, 103 ,15.3+2,111.45,0
K, 104 ,14.95+2,120.2,0
K, 105 ,24.35,118.9,0
K, 106 ,28.9,111.75,0
K, 107 ,32.15,98.15,0
K, 108 ,34.75,78.4,0
K, 109 ,32.15,62.55,0
K, 110 ,26+2,47.95,0
K, 111 ,24+2,31.45,0
K, 112 ,22+2,15.25,0
```

```
K, 113 ,22+2,-5.2,0
K, 114 ,22+2,-20.95,0
K, 115 ,22+2,-30,0
K, 116 ,18.85+2,122.45-2,0
K, 117 ,-18,-30,0
K, 118 ,-18,-12.9,0
K, 119 ,-18,5.25,0
K, 120 ,-18,25.95,0
K, 121 ,-18.45,35.05,0
K, 122 ,-19.45,44,0
K, 123 ,-21.65,52.75,0
K, 124 ,-24.25,60.55,0
K, 125 ,-7.25,53.25,0
K, 126 ,-7,48.8,0
K, 127 ,-7.8,40.8,0
K, 128 ,-7.45,32.8,0
K, 129 ,-7.15,22.15,0
K, 130 ,-7.35,1.25,0
K, 131 ,-7.35,-7.2,0
K, 132 ,-7.35,-17.85,0
K, 133 ,-7.35,-27.2,0
K, 134 ,-7,-30,0
K, 135 ,22-2,95.45,0
K, 136 ,25.75-3,87,0
K, 137 ,27.55-5,74.15,0
K, 138 ,24-4,63.9,0
K, 139 ,19.1-2,52.8,0
K, 140 ,16-2,31.45,0
K, 141 ,13.75,16.8,0
K, 142 ,12.45,-6.3,0
K, 143 ,12.45,-20.3,0
K, 144 ,12.85,-25.85,0
K, 145 ,12.25,-30,0
!*************
K, 146 ,-5.4,40.95,0
K, 147 ,-5.65,26.7,0
K, 148 ,-5.65,12.1,0
K, 149 ,-5.3,2.4,0
K, 150 ,-5.3,-8.95,0
K, 151 ,10.45,44.85,0
K, 152 ,10.05,38.65,0
K, 153 ,10.75,23.75,0
K, 154 ,10.1,13.1,0
```

```
K, 155 ,9.8,4,0
LSTR,    30,    1
LSTR,    29,   28
LSTR,     2,    3
LSTR,    28,   27
LSTR,     3,    4
LSTR,     6,    7
LSTR,     7,    8
LSTR,     8,    9
LSTR,    23,   22
LSTR,    22,   21
LSTR,    21,   20
LSTR,    19,   18
LSTR,    10,   11
LSTR,    13,   14
LSTR,    14,   35
LSTR,    35,   15
LSTR,    15,   16
LSTR,    16,   35
LSTR,    35,   13
LSTR,     7,   22
!*********************************************
LARC,30,29,32,1.5,
LARC,1,2,31,1.5,
LARC,20,19,33,1,
LARC,10,9,34,1,
!*********************************************
FLST,3,3,3
FITEM,3,16
FITEM,3,17
FITEM,3,18
BSPLIN, ,P51X
FLST,3,3,3
FITEM,3,13
FITEM,3,12
FITEM,3,11
BSPLIN, ,P51X
FLST,3,3,3
FITEM,3,4
FITEM,3,5
FITEM,3,6
BSPLIN, ,P51X
FLST,3,5,3
```

```
FITEM,3,27
FITEM,3,26
FITEM,3,25
FITEM,3,24
FITEM,3,23
BSPLIN, ,P51X
!*********************************************
LSTR,    21,    101
LSTR,     8,    124
LSTR,   117,    134
LSTR,   145,    115
FLST,3,4,3
FITEM,3,101
FITEM,3,102
FITEM,3,103
FITEM,3,104
BSPLIN, ,P51X
FLST,3,13,3
FITEM,3,104
FITEM,3,116
FITEM,3,105
FITEM,3,106
FITEM,3,107
FITEM,3,108
FITEM,3,109
FITEM,3,110
FITEM,3,111
FITEM,3,112
FITEM,3,113
FITEM,3,114
FITEM,3,115
BSPLIN, ,P51X
FLST,3,12,3
FITEM,3,101
FITEM,3,135
FITEM,3,136
FITEM,3,137
FITEM,3,138
FITEM,3,139
FITEM,3,140
FITEM,3,141
FITEM,3,142
FITEM,3,143
```

```
FITEM,3,144
FITEM,3,145
BSPLIN, ,P51X
GPLOT
FLST,3,11,3
FITEM,3,6
FITEM,3,125
FITEM,3,126
FITEM,3,127
FITEM,3,128
FITEM,3,129
FITEM,3,130
FITEM,3,131
FITEM,3,132
FITEM,3,133
FITEM,3,134
BSPLIN, ,P51X
FLST,3,8,3
FITEM,3,124
FITEM,3,123
FITEM,3,122
FITEM,3,121
FITEM,3,120
FITEM,3,119
FITEM,3,118
FITEM,3,117
BSPLIN, ,P51X
FLST,3,7,3
FITEM,3,4
FITEM,3,146
FITEM,3,147
FITEM,3,148
FITEM,3,149
FITEM,3,150
FITEM,3,134
BSPLIN, ,P51X
FLST,3,7,3
FITEM,3,27
FITEM,3,151
FITEM,3,152
FITEM,3,153
FITEM,3,154
FITEM,3,155
```

```
FITEM,3,145
BSPLIN, ,P51X
FLST,2,3,4
FITEM,2,18
FITEM,2,17
FITEM,2,16
AL,P51X
FLST,2,13,4
FITEM,2,20
FITEM,2,10
FITEM,2,11
FITEM,2,23
FITEM,2,12
FITEM,2,25
FITEM,2,18
FITEM,2,19
FITEM,2,26
FITEM,2,13
FITEM,2,24
FITEM,2,8
FITEM,2,7
AL,P51X
FLST,2,3,4
FITEM,2,19
FITEM,2,15
FITEM,2,14
AL,P51X
FLST,2,12,4
FITEM,2,20
FITEM,2,6
FITEM,2,27
FITEM,2,5
FITEM,2,3
FITEM,2,22
FITEM,2,1
FITEM,2,21
FITEM,2,2
FITEM,2,4
FITEM,2,28
FITEM,2,9
AL,P51X
FLST,2,4,4
FITEM,2,32
```

```
FITEM,2,34
FITEM,2,33
FITEM,2,35
AL,P51X
FLST,2,6,4
FITEM,2,35
FITEM,2,29
FITEM,2,10
FITEM,2,9
FITEM,2,28
FITEM,2,39
AL,P51X
FLST,2,3,4
FITEM,2,38
FITEM,2,27
FITEM,2,36
AL,P51X
FLST,2,6,4
FITEM,2,31
FITEM,2,36
FITEM,2,6
FITEM,2,7
FITEM,2,30
FITEM,2,37
AL,P51X
!*** Element Type Definition
ET,1,PLANE82
ET,2,PLANE82
ET,3,PLANE82
!*** Material Properities
MPTEMP,,,,,,,,
MPTEMP,1,0
MPDATA,EX,1,,EM
MPDATA,PRXY,1,,NUM
MPTEMP,,,,,,,,
MPTEMP,1,0
MPDATA,EX,2,,ECOR
MPDATA,PRXY,2,,NUCOR
MPTEMP,,,,,,,,
MPTEMP,1,0
MPDATA,EX,3,,ECAN
MPDATA,PRXY,3,,NUCAN
!*** Meshing Creation
```

```
TYPE,   1
MAT,       1
REAL,
ESYS,       0
SECNUM,
!************
SMRT,6
SMRT,1
MSHAPE,0,2D
MSHKEY,0
!************
FLST,5,4,5,ORDE,2
FITEM,5,1
FITEM,5,-4
CM,_Y,AREA
ASEL, , , ,P51X
CM,_Y1,AREA
CHKMSH,'AREA'
CMSEL,S,_Y
AMESH,_Y1
CMDELE,_Y
CMDELE,_Y1
CMDELE,_Y2
/UI,MESH,OFF
GPLOT
TYPE,   2
MAT,       2
REAL,
ESYS,       0
SECNUM,
!************
FLST,5,2,5,ORDE,2
FITEM,5,5
FITEM,5,8
CM,_Y,AREA
ASEL, , , ,P51X
CM,_Y1,AREA
CHKMSH,'AREA'
CMSEL,S,_Y
AMESH,_Y1
CMDELE,_Y
CMDELE,_Y1
CMDELE,_Y2
```

```
/UI,MESH,OFF
GPLOT
TYPE,  3
MAT,     3
REAL,
ESYS,     0
SECNUM,
!************
FLST,5,2,5,ORDE,2
FITEM,5,6
FITEM,5,-7
CM,_Y,AREA
ASEL, , , ,P51X
CM,_Y1,AREA
CHKMSH,'AREA'
CMSEL,S,_Y
AMESH,_Y1
CMDELE,_Y
CMDELE,_Y1
CMDELE,_Y2
/UI,MESH,OFF
FINISH
!*****************
!*** Solution Process
!*****************
/SOL
!*** Analysis Type Definition
ANTYPE,0
!*** Boundary Condition Definition
FLST,2,1,3,ORDE,1
FITEM,2,15
/GO
FK,P51X,FX,FB1X
FLST,2,1,3,ORDE,1
FITEM,2,15
/GO
FK,P51X,FY,-FB1Y
FLST,2,1,3,ORDE,1
FITEM,2,35
/GO
FK,P51X,FX,FB2X
FLST,2,1,3,ORDE,1
```

```
FITEM,2,35
/GO
FK,P51X,FY,-FB2Y
FLST,2,1,3,ORDE,1
FITEM,2,14
/GO
FK,P51X,FX,FB3X
FLST,2,1,3,ORDE,1
FITEM,2,14
/GO
FK,P51X,FY,-FB3Y
FLST,2,1,3,ORDE,1
FITEM,2,104
/GO
FK,P51X,FX,-FM1X
FLST,2,1,3,ORDE,1
FITEM,2,104
/GO
FK,P51X,FY,FM1Y
!************
FLST,2,2,4,ORDE,2
FITEM,2,31
FITEM,2,-32
/GO
DL,P51X, ,ALL,
/STATUS,SOLU
SOLVE
FINISH
!***************
!*** Postprocessor
!***************
/POST1
/EFACET,1
PLNSOL, S,EQV, 0,1.0
!*** Volume Evaluation
AVPRIN,0, ,
ETABLE, ,VOLU,
SSUM
*GET,VOLUME,SSUM, ,ITEM,VOLU
!*** Maximum Stress Evaluation
FLST,5,2,5,ORDE,2
FITEM,5,5
```

```
FITEM,5,8
ASEL,S, , ,P51X
NSLA,R,1
NSORT,S,EQV,0,0, ,0
*GET,SMAX1,SORT,,MAX
!*** Maximum Stress Evaluation
AVPRIN,0, ,
NSEL,ALL
FLST,5,2,5,ORDE,2
FITEM,5,6
FITEM,5,-7
ASEL,S, , ,P51X
NSLA,R,1
NSORT,S,EQV,0,0, ,0
*GET,SMAX2,SORT,,MAX
!*** Maximum Stress Evaluation
AVPRIN,0, ,
NSEL,ALL
FLST,5,4,5,ORDE,2
FITEM,5,1
FITEM,5,-4
ASEL,S, , ,P51X
NSLA,R,1
NSORT,S,EQV,0,0, ,0
*GET,SMAXM,SORT,,MAX
!*** Minimum Stress Evaluation
AVPRIN,0, ,
NSEL,ALL
FLST,5,2,5,ORDE,2
FITEM,5,5
FITEM,5,8
ASEL,S, , ,P51X
NSLA,R,1
NSORT,S,EQV,0,0, ,0
*GET,SMIN1,SORT,,MIN
!*** Minimum Stress Evaluation
AVPRIN,0, ,
NSEL,ALL
FLST,5,2,5,ORDE,2
FITEM,5,6
FITEM,5,-7
ASEL,S, , ,P51X
```

```
NSLA,R,1
NSORT,S,EQV,0,0, ,0
*GET,SMIN2,SORT,,MIN
!*** Minimum Stress Evaluation
AVPRIN,0, ,
NSEL,ALL
FLST,5,4,5,ORDE,2
FITEM,5,1
FITEM,5,-4
ASEL,S, , ,P51X
NSLA,R,1
NSORT,S,EQV,0,0, ,0
*GET,SMINM,SORT,,MIN
!*** Limite State Evaluation
DIF_SMAX1=YS_COR-SMAX1
DIF_SMAX2=YS_CAN-SMAX2
DIF_SMIN1=SMIN1-ST_COR
DIF_SMIN2=SMIN2-ST_CAN
!*** Beta Evaluation
*SET,BATOT,1/BETA
!***End ANSYS Code
!*************************************************
```

Appendix 4

Probability of Failure/Reliability Index

Using a FORM approximation, the probability of failure is simply calculated by:

$$P_f \approx \Phi(-\beta) \quad \text{or} \quad \beta \approx -\Phi^{-1}(P_f)\text{, with } \Phi(Z) = \frac{1}{\sqrt{2\pi}} \int_{-\infty}^{Z} e^{-\frac{z^2}{2}} dz,$$

β	P_f	β	P_f	β	P_f
3.00	0.001358	2.00	0.022853	1.00	0.159127
2.95	0.001598	1.95	0.025702	0.95	0.171553
2.90	0.001877	1.90	0.028842	0.90	0.184581
2.85	0.002199	1.85	0.032294	0.85	0.198207
2.80	0.002570	1.80	0.036081	0.80	0.212424
2.75	0.002997	1.75	0.040224	0.75	0.227219
2.70	0.003486	1.70	0.044749	0.70	0.242578
2.65	0.004047	1.65	0.049668	0.65	0.258482
2.60	0.004687	1.60	0.055013	0.60	0.274910
2.55	0.005415	1.55	0.060802	0.55	0.291837
2.50	0.006243	1.50	0.067057	0.50	0.309233
2.45	0.007180	1.45	0.073798	0.45	0.327069
2.40	0.008240	1.40	0.081046	0.40	0.345308
2.35	0.009434	1.35	0.088818	0.35	0.363914
2.30	0.010778	1.30	0.097132	0.30	0.382846
2.25	0.012285	1.25	0.106004	0.25	0.402062
2.20	0.013971	1.20	0.115447	0.20	0.421519
2.15	0.015853	1.15	0.125472	0.15	0.441169
2.10	0.017948	1.10	0.136090	0.10	0.460964
2.05	0.020275	1.05	0.147307	0.05	0.480857
				0.00	0.500798

Bibliography

[ALE 13] ALEIXO I., VALE A.C., LÚCIO M. *et al.*, "Method for the evaluation of femoral head trabecular bone compressive properties", *Materials Science Forum*, pp. 3–8, 2013.

[AOU 10] AOUN M., Elaboration et validation d'un modèle de l'Articulation Temporo-Mandibulaire par éléments finis, PhD Thesis, University of Bordeaux 1, 2010.

[AOU 11] AOUN M., MESNARD M., MORLIER J. *et al.* "Elaboration and validation of a 2D finite element model of the temporomandibular joint using magnetic resonance images", *Russian Journal of Biomechanics*, vol. 1, no. 51, pp. 21–29, 2011.

[BER 95] BERTILSSON O., STROM D., "Anatomic and functional lateral pterygoid muscle research", *Orofacial Pain*, vol. 9, pp. 17–23, 1995.

[BIL 00] BILGIN T., SÜLÜN T., ERGIN U. *et al.*, "Evaluation of the slope of the articular eminence and the transverse angle of the glenoid fossa in an anatolian population", *Journal of Craniomandibular Practice*, vol. 18, no. 3, pp. 220–227, 2000.

[CAI 91] CAIX Ph., CARLES J., L'articulation temporo-mandibulaire et son vrai-faux ménisque, Laboratoire d'Anatomie, Faculté de Médecine, Amiens, 1991.

[CAI 00] CAIX Ph., CARLES J., Anatomie fonctionnelle, Laboratoire d'Anatomie Médico-Fonctionnelle et Chirurgicale, University of Bordeaux 2, 2000.

[CAI 02] CAIX Ph., CARLES J., Le système axial: tête, rachis, tronc, Laboratoire d'Anatomie Médico-Fonctionnelle et Chirurgicale, University of Bordeaux 2, 2000.

[CHI 05] CHIARI L., DELLA GROCE U., LEARDINI A. *et al.*, "Human movement analysis using stereophotogrammetry, Part 2: instrumental errors", *Gait and Posture*, vol. 21, pp. 197–211, 2005.

[COR 76] CORNELL C.A., "Bounds on the reliability of structural systems", *J. struct. div.*, ASCE, vol. 93, no. 1, pp. 171–200, 1967.

[COT 00] COTTEN A., DEMONDION X., "Développement et vieillissement du disque intervertébral lombaire", *Revue du Rhumatisme*, vol. 67, no. 4, pp. 207–213, 2000.

[COX 03] COX T., KOHN M.W., IMPELLUSO T., "Computerized analysis of resorbable polymer plates and screws for the rigid fixation of mandibular angle fractures", *Journal of Oral and Maxillofacial Surgery*, vol. 61, pp. 481–487, 2003.

[DEL 05] DELLA CROCE U., LEARDINI A., CHIARI L. *et al.*, "Human movement analysis using stereophotogrammetry, Part 4: assessment of anatomical landmark misplacement and its effects on joint kinematics", *Gait and Posture*, vol. 21, no. 2, pp. 226–237, 2005.

[DOB 04] DOBLARE M., GARCIA J.M., GOMEZ M.J., "Modeling bone tissue fracture and healing: a review", *Engineering Fracture Mechanics*, vol. 71, pp. 1809–1840, 2004.

[ELH 11] EL HAMI A., RADI B., *Fiabilité et optimisation des systèmes: Théorie et applications*, Ellipses Editions, Paris, 2011.

[FAR 63] FARMER F.R., "Siting criteria, a new approach", *Atom*, vol. 128, pp. 152–170, 1967.

[FIN 83] FINET A., *Le Code d'Hammourabi*, Éditions du Cerf, Paris, 1983.

[FOR 82] FORTÉ M.C., Le bâillement, Laboratoire d'Anatomie, University of Bordeaux 2, 1982.

[GEN 10] GENNISSON J., DEFFIEUX T., MACE E. *et al.*, "Viscoelastic and an-isotropic mechanical properties of in vivo muscle tissue assessed by supersonic shear imaging", *Ultrasound in Medicine & Biology*, vol. 36, pp. 789–801, 2010.

[GIB 76] GIBBONS D.F., "Biomedical materials", in FLEMING D.G., FEINBERG B.N. (eds), *Handbook of Engineering in Medicine and Biology*, CRC Press, Boca Raton, Florida, 1976.

[GOR 09] GORTON G.E., HEBERT D., GANNOTTI M.E., "Assessment of the kinematic variability among 12 motion analysis laboratories", *Gait and Posture*, vol. 29, no. 3, pp. 398–402, 2009.

[GRA 98] GRANDHI R.V., WANG L., "Reliability-based structural optimization using improved two-point adaptive nonlinear approximations", *Journal of Finite Elements in Analysis and Design*, vol. 29, pp. 35–48, 1998.

[GRO 01] GROSS S., ABEL E.W., "A finite element analysis of hollow stemmed hip prostheses as a means of reducing stress shielding of the femur", *Journal of Biomechanics*, vol. 34, pp. 995–1003, 2001.

[HAN 94] HANNAM A.G., MCMILLAN A., "Internal organisation in the human jaw muscles", *Critical Reviews in Oral Biology and Medicine*, vol. 5, pp. 55–89, 1994.

[HAS 76] HASOFER A.M., LIND N.C., "An exact and invariant first order reliability format", *Journal of Engineering Mechanics*, ASCE, EM1, vol. 100, pp. 111–121, 1976.

[HEY 95] HEYLINGS D., NIELSEN I.L., MC NEILL C., "The lateral pterygoid muscle and the temporomandibular disc", *Orofacial Pain*, vol. 9, pp. 9–16, 1995.

[HUI 87] HUISKES R., WEINANS H., GROOTENBOER H.J. *et al.*, "Adaptive bone-remodeling theory applied to prosthetic-design analysis", *Journal of Biomechanics*, vol. 20, pp. 1135–1150, 1987.

[JAL 06] JALLOUFF S., Approche probabiliste du dimensionnement contre le risque de rupture, PhD Thesis, University of Paul Verlaine de Metz – University of Aleppo, 2006.

[KAD 07] KADRY S., Couplage éléments finis et méthode de transformation probabiliste, PhD Thesis, University of Blaise Pascal – Clermont II, 2007.

[KAN 10] KANER T., SASANI M., OKTENOGLU T. *et al.*, "Dynamic stabilization of the spine: a new classification system", *Turkish Neurosurgery*, vol. 20, no. 2, pp. 205–215, 2010.

[KAP 07] Kapandji A.I. *Anatomie fonctionnelle III - Tête et rachis*, Maloine, 6th edition, 2007.

[KAP 11] KAPANDJI A.I., *Q'est ce que la Biomécanique?*, Sauramps Medical, 2001.

[KAY 08] KAYABASI O., EKICI B., "Probabilistic design of a newly designed cemented hip prosthesis using finite element method", *Journal of Materials and Design*, vol. 29, no. 5, pp. 963–971, 2008.

[KEA 01] KEAVENY T.M., MORGAN E.F., NIEBUR G.L. *et al.*, "Biomechanics of trabecular bone", *Annual Review of Biomedical Engineering*, vol. 3, pp. 307–333, 2001.

[KEL 94] KELLER T.S., "Predicting the compressive mechanical behavior of bone", *Journal of Biomechanics*, vol. 27, pp. 1159–1168, 1994.

[KEY 94] KEYAK J.H., LEE I., SKINNER H.B., "Correlations between orthogonal mechanical properties and density of trabecular bone: use of different densitometric measures", *J. Biomech. Mater. Res.*, vol. 28, pp. 1329–1336, 1994.

[KHA 03] KHARMANDA G., Optimization and CAD of Reliable Structures, PhD thesis, Blaise Pascal University, 23 June 2003.

[KHA 08] KHARMANDA G., On reliability-based structural optimization: Methodologies and applications, H.D.R. thesis, University of Rouen, 2008.

[KHA 09] KHARMANDA G., SHARABATY S., IBRAHIM M.H. *et al.*, "Reliability-based design optimization using semi-numerical methods for different engineering application", *International Journal of CAD/CAM*, vol. 9, pp. 1–16, 2009.

[KHA 11] KHARMANDA G., *Computer-Aided Design*, Aleppo University Publications, Aleppo, 2011.

[KHA 14] KHARMANDA G., KHARMA M.Y., RISTINMAA M. *et al.*, "Structural optimization of mini-plates in fixation of human mandible fractures", *27th Nordic Seminar on Computational Mechanics – NSCM-27*, Stockholm, October 2014.

[KHA 15a] KHARMANDA G., "Reliability analysis for cementless hip prosthesis using a new optimized formulation of yield stress against elasticity modulus relationship", *Materials and Design*, vol. 65, pp. 496–504, 2015.

[KHA 15b] KHARMANDA G., SHORKY A., "Investigation and validation of a new optimized yield stress/elasticity modulus formulation for bone fracture identification", *ICMCSF*, Lille, France, May 17-22, 2015.

[KHA 16a] KHARMANDA G., "Integration of multi-objective structural optimization into cementless hip prosthesis design: improved Austin-Moore model", *Computer Methods in Biomechanics and Biomedical Engineering*, vol. 19, no. 14, pp. 1557–1566, 2016.

[KHA 16b] KHARMANDA G., KHARMA M.Y., "Evaluating the effect of minimizing screws on stabilization of symphysis mandibular fracture by 3D finite element analysis", *Journal of Maxillofacial and Oral Surgery*, 21 April, 2016.

[KHA 16c] KHARMANDA G., IBRAHIM H., EL HAMI A., "Reliability analysis of mini-plate stabilization used in human fractured mandibles", *ESB Annual Congress*, 10 July 2016.

[KOR 07] KORKMAZ H.H., "Evaluation of different mini-plates in fixation of fractured human mandible with the finite element method", *Oral Surgery, Oral Medicine, Oral Pathology and Oral Radiology*, vol. 103, pp. e1–e13, 2007.

[KUB 88] KUBEIN-MEESENBURG D., NÄGER H., KLAM B., "The biomechanical relation between incisal and condylar guidance in man", *Journal of Biomechamics*, vol. 21, no. 12, pp. 997–1009, 1988.

[KUM 13] KUMAR S.T., SARAF S., DEVI S.P., "Evaluation of bite force after open reduction and internal fixation using microplates", *Journal of Dentistry*, Tehran University of Medical Sciences, vol. 10, no. 5, pp. 466–477, 2013.

[LAS 08] LAS CASAS E.B., FERREIRA P.C., CIMINI CA J.R. *et al.*, "Comparative 3D finite element stress analysis of straight and angled wedge-shaped implant designs", *The International Journal of Oral & Maxillofacial Implants*, vol. 23, no. 2, pp. 215–225, 2008.

[LAU 07] LAU S., LAM K.S. "Lumbar stabilisation techniques", *Current Orthopaedics*, vol. 21, pp. 25–39, 2007.

[LEA 05] LEARDINI A., CHIARI L., DELLA CROCE U. *et al.*, "Human movement analysis using stereophotogrammetry, Part 3: Soft tissue artifact assessment and compensation", *Gait and Posture*, vol. 21, no. 2, pp. 212–225, 2005.

[LEE 07] LEE A.C., MAGEE D.J., ZACHAZEWSKI J.E. *et al.*, "Injury, inflammation, and repair: tissue mechanics, the healing process, and their impact on the musculoskeletal system", in MAGEE D.J., ZACHAZEWSKI J.E., QUILLEN W.S. (eds), *Scientific Foundations and Principles of Practice in Musculoskeletal Rehabilitation*, Saunders, Philadelphia, 2007.

[LEM 07] LEMAIRE J.P., DAVID T., MORENO P., *Complications des prothèses de disque intervertébrales. Alternatives à l'arthrodèse lombaire et lombosacrée*, Elsevier, 2007.

[LEM 09] LEMAIRE M., *Structural Reliability*, ISTE, London and John Wiley & Sons, New York, 2009.

[LIT 97] LITTRE E., *Dictionnaire Britanica France*, Versailles, 1997.

[MAR 74] MARCOVICI C., VIGIER M.G., *Utilisation des techniques de fiabilité en mécanique*, Tec & Doc, Paris, 1974.

[MES 11] MESNARD M., RAMOS A., BALLU A. *et al.*, "Biomechanical analysis comparing natural and alloplastic temporomandibular joint replacement using a finite element model", *Journal of Oral and Maxillofacial Surgery*, vol. 69, no. 4, pp. 1008–1017, 2011.

[MOD 89] MODSCHIEDLER T., "La place de la prothèse dans la chirurgie reconstructive de l'articulation temporo-mandibulaire", *Sciences Médicales*, University of Bordeaux 2, 1989.

[NAG 02] NAGASAO T., KOBAYASHI M., TSUCHIYA Y. *et al.*, "Finite element analysis of the stresses around endosseous implants in various reconstructed mandibular models", *Journal of Cranio-Maxillo-Facial Surgery*, vol. 30, pp. 170–177, 2002.

[PAL 07] PALOMBI O., Le segment mobile. Anatomie de la colonne vertébrale, Faculté de Médecine de Grenoble, 2007.

[PET 05] PETRIE C.S., WILLIAMS J.L., "Comparative evaluation of implant designs: influence of diameter, length, and taper on strains in the alveolar crest. A three-dimensional finite element analysis", *Journal of Clinical Oral Implants Research*, vol. 16, pp. 486–494, 2005.

[POU 99] POURROY F., "Apport des éléments finis à la conception mécanique", *Techniques de l'Ingénieur*, Référence AG2530, 10 April 1999.

[RAM 11] RAMOS A., BALLU A., MESNARD M. *et al.*, "Numerical and experimental models of the mandible", *Experimental Mechanics*, vol. 51, no. 7, pp. 1053–1059, 2011.

[RAM 14] RAMOS A., MARQUES H., MESNARD M., "The effect of mechanical properties of bone in the mandible, a numerical case study", *Advances in Biomechanics and Applications*, vol. 1, no. 1, pp. 067–076, 2014.

[RAN 04] RANNOU F., MAYOUX-BENHAMOU M.A., POIRAUDEAU S. *et al.*, "Disque intervertébral et structures voisines des la colonne lombaire: anatomie, biologie, physiologie et biomécanique", *EMC-Rhumatologie Orthopédie*, vol. 1, pp. 487–507, 2004.

[REI 75] REILLY D.T., BURSTEIN A.H., "The elastic and ultimate properties of compact bone tissue", *Journal of Biomechanics*, vol. 8, pp. 393–405, 1975.

[SCH 74] SCHOENFELD C.M., LAUTENSCHLAGER E.P., MEYER P.R., "Mechanical properties of human cancellous bone in the femoral head", *Journal of Medical and Biological Engineering*, vol. 12, pp. 313–317, 1974.

[SER 11] SERHAN H., MHATRE D., DEFOSSEZ H. *et al.*, "Motion-preserving technologies for degenerative lumbar spine: the past, present, and future horizons", *SAS Journal*, vol. 5, pp. 75–89, 2011.

[SHO 15] SHOKRY A. On the determination of diffusion coefficients in bone, PhD thesis, Lund University, 2015.

[STE 00] STEINWENDER G., SARAPH V., SCHEIBER S. *et al.*, "Intrasubject repeatability of gait analysis data in normal and spastic children", *Clinical Biomechanics* (Bristol, Avon), vol. 15, no. 2, pp. 134–139, 2000.

[TEB 05] TEBBI O., Estimation des lois de fiabilité en mécanique par les essais accélérés, PhD Thesis, École doctorale d'Angers, 2005.

[TSU 02] TSUBOTA K., ADACHI T., TOMITA Y., "Functional adaptation of cancellous bone in human proximal femur predicted by trabecular surface remodeling simulation toward uniform stress state", *Journal of Biomechanics*, vol. 35, pp. 1541–1551, 2002.

[VIT 07] VITAL J.M. *Processus de Dégénérescence Lombaire. Alternatives à L'Arthrodèse Lombaire et Lombosacrée*, Elsevier, 2007.

[WEI 03] WEISSE B., ZAHNER M., WEBER W. *et al.*, "Improvement of the reliability of ceramic hip joint implants", *Journal of Biomechanics*, vol. 36, no. 11, 2003.

[WOL 86] WOLFF J., *The Law of Bone Remodeling*, Springer, Berlin, 1986.

[YAV 08] YAVUZER G., OKEN O., ELHAN A. *et al.*, "Repeatability of lower limb three-dimensional kinematics in patients with stroke", *Gait and Posture*, vol. 27, no. 1, pp. 31–35, 2008.

Index

O, P

R, S, U

Other titles from

ISTE

in

Mechanical Engineering and Solid Mechanics

2016

BOREL Michel, VÉNIZÉLOS Georges
Movement Equations 1: Location, Kinematics and Kinetics

BOYARD Nicolas
Heat Transfer in Polymer Composite Materials

CARDON Alain, ITMI Mhamed
New Autonomous Systems
(Reliability of Multiphysical Systems Set - Volume 1)

DAHOO Pierre Richard, POUGNET Philippe, EL HAMI Abdelkhalak
Nanometer-scale Defect Detection Using Polarized Light
(Reliability of Multiphysical Systems Set – Volume 2)

DE SAXCÉ Géry
Galilean Mechanics and Thermodynamics of Continua

DORMIEUX Luc, KONDO Djimédo
Micromechanics of Fracture and Damage (Micromechanics Set - Volume 1)

MOLIMARD Jérôme
Experimental Mechanics of Solids and Structures

RADI Bouchaib, EL HAMI Abdelkhalak
Material Forming Processes: Simulation, Drawing, Hydroforming and Additive Manufacturing

2015

KARLIČIĆ Danilo, MURMU Tony, ADHIKARI Sondipon, MCCARTHY Michael
Non-local Structural Mechanics

SAB Karam, LEBÉE Arthur
Homogenization of Heterogeneous Thin and Thick Plates

2014

ATANACKOVIC M. Teodor, PILIPOVIC Stevan, STANKOVIC Bogoljub, ZORICA Dusan
Fractional Calculus with Applications in Mechanics: Vibrations and Diffusion Processes

ATANACKOVIC M. Teodor, PILIPOVIC Stevan, STANKOVIC Bogoljub, ZORICA Dusan
Fractional Calculus with Applications in Mechanics: Wave Propagation, Impact and Variational Principles

CIBLAC Thierry, MOREL Jean-Claude
Sustainable Masonry: Stability and Behavior of Structures

ILANKO Sinniah, MONTERRUBIO Luis E., MOCHIDA Yusuke
The Rayleigh−Ritz Method for Structural Analysis

LALANNE Christian
Mechanical Vibration and Shock Analysis – 5-volume series – 3rd edition
Sinusoidal Vibration – volume 1
Mechanical Shock – volume 2
Random Vibration – volume 3
Fatigue Damage – volume 4
Specification Development – volume 5

LEMAIRE Maurice
Uncertainty and Mechanics

2013

ADHIKARI Sondipon
Structural Dynamic Analysis with Generalized Damping Models: Analysis

ADHIKARI Sondipon
Structural Dynamic Analysis with Generalized Damping Models: Identification

BAILLY Patrice
Materials and Structures under Shock and Impact

BASTIEN Jérôme, BERNARDIN Frédéric, LAMARQUE Claude-Henri
Non-smooth Deterministic or Stochastic Discrete Dynamical Systems: Applications to Models with Friction or Impact

EL HAMI Abdelkhalak, BOUCHAIB Radi
Uncertainty and Optimization in Structural Mechanics

KIRILLOV Oleg N., PELINOVSKY Dmitry E.
Nonlinear Physical Systems: Spectral Analysis, Stability and Bifurcations

LUONGO Angelo, ZULLI Daniele
Mathematical Models of Beams and Cables

SALENÇON Jean
Yield Design

2012

DAVIM J. Paulo
Mechanical Engineering Education

DUPEUX Michel, BRACCINI Muriel
Mechanics of Solid Interfaces

ELISHAKOFF Isaac *et al.*
Carbon Nanotubes and Nanosensors: Vibration, Buckling and Ballistic Impact

GRÉDIAC Michel, HILD François
Full-Field Measurements and Identification in Solid Mechanics

GROUS Ammar
Fracture Mechanics – 3-volume series
Analysis of Reliability and Quality Control – volume 1
Applied Reliability – volume 2
Applied Quality Control – volume 3

RECHO Naman
Fracture Mechanics and Crack Growth

2011

KRYSINSKI Tomasz, MALBURET François
Mechanical Instability

SOUSTELLE Michel
An Introduction to Chemical Kinetics

2010

BREITKOPF Piotr, FILOMENO COELHO Rajan
Multidisciplinary Design Optimization in Computational Mechanics

DAVIM J. Paulo
Biotribolgy

PAULTRE Patrick
Dynamics of Structures

SOUSTELLE Michel
Handbook of Heterogenous Kinetics

2009

BERLIOZ Alain, TROMPETTE Philippe
Solid Mechanics using the Finite Element Method

LEMAIRE Maurice
Structural Reliability

2007

GIRARD Alain, ROY Nicolas
Structural Dynamics in Industry

GUINEBRETIÈRE René
X-ray Diffraction by Polycrystalline Materials

KRYSINSKI Tomasz, MALBURET François
Mechanical Vibrations

KUNDU Tribikram
Advanced Ultrasonic Methods for Material and Structure Inspection

SIH George C. *et al.*
Particle and Continuum Aspects of Mesomechanics

www.ingramcontent.com/pod-product-compliance
Lightning Source LLC
Chambersburg PA
CBHW071300180926
51613CB00001B/79
* 9 7 8 1 7 8 6 3 0 0 2 4 9 *